BRAIN-FRIENDLY ASSESSMENTS

What They Are and How to Use Them

BRAIN-FRIENDLY ASSESSMENTS

What They Are and How to Use Them

David A. Sousa

LearningSciencesInternational

1400 Centrepark Blvd, Suite 1000
West Palm Beach, FL 33401
717-845-6300

email: pub@learningsciences.com
learningsciences.com

Printed in the United States of America

20 19 18 17 16 15   2 3 4

FSC
www.fsc.org
MIX
Paper from
responsible sources
FSC® C005010

Publisher's Cataloging-in-Publication Data
provided by Five Rainbows Services

Sousa, David A.
Brain-friendly assessments : what they are and how to use them / David A. Sousa.
pages cm
ISBN: 978-1-941112-21-2 (pbk.)
1. Educational tests and measurements. 2. Brain. 3. Learning—Physiological aspects. 4. Neurosciences. 5. Cognitive learning.   I. Title.
LB3051 .S68 2014
371.26—dc23
[2014955749]

Table of Contents

Acknowledgments

Learning Sciences International would like to thank the following reviewers:

Steve Atwater, PhD
Superintendent
Kenai Peninsula Borough School
   District
Soldotna, Alaska

Darren T. Guido, EdD
Supervisor of Instruction
Capital School District
Dover, Delaware

Dr. C. J. Huff
Superintendent
Joplin Schools
Joplin, Missouri

Brandi Leggett
2014 Regional Kansas Teacher
   of the Year
Prairie Ridge Elementary School
Shawnee, Kansas

Alana M. Margeson
2012 Maine Teacher of the Year
Caribou High School
Caribou, Maine

Dr. Tim Richard
Principal
Poston Butte High School
San Tan Valley, Arizona

About the Author

DR. DAVID A. SOUSA is an international consultant in educational neuroscience and author of sixteen books that suggest ways educators and parents can translate current brain research into strategies for improving learning. A member of the Cognitive Neuroscience Society, he has conducted workshops in hundreds of school districts on brain research, instructional skills, and science education at the preK–12 and university levels. He has made presentations to more than two hundred thousand educators at national conventions of educational organizations and to regional and local school districts across the United States, Canada, Europe, Australia, New Zealand, and Asia.

Dr. Sousa has a bachelor's degree in chemistry from Bridgewater State University in Massachusetts, a master of arts in teaching degree in science from Harvard University, and a doctorate from Rutgers University. His teaching experience covers all levels. He has taught senior high school science, and served as a K–12 director of science, supervisor of instruction, and district superintendent in New Jersey schools. He was an adjunct professor of education at Seton Hall University for ten years and a visiting lecturer at Rutgers University.

Prior to his career in New Jersey, Dr. Sousa taught at the American School of Paris (France) and served for five years as a foreign service officer and science advisor at the US diplomatic missions in Geneva (Switzerland) and Vienna (Austria).

Dr. Sousa has edited science books and published dozens of articles in leading journals on professional development, science education, and educational research. His most popular books for educators include *How the Brain Learns*, now in its fourth edition; *How the Special Needs Brain Learns*, second edition; *How the Gifted Brain Learns*; *How the Brain Learns to Read*, second edition; *How the Brain Influences Behavior*; *How the ELL Brain Learns*; *Differentiation and the Brain* (with Carol Tomlinson); and *How the Brain Learns Mathematics*, which was selected by the Independent Book Publishers' Association as one of the best professional-development books of 2008 and is now being revised for a second edition. *The Leadership Brain* suggests ways for educators to lead today's schools more effectively. His books have been published in French, Spanish, Chinese, Arabic, Korean, Russian, and several other languages. His book *Brainwork: The Neuroscience Behind How We Lead Others* is written for business and organizational leaders.

Dr. Sousa is past president of the National Staff Development Council (now called Learning Forward). He has received numerous awards from professional associations, school districts, and educational foundations for his commitment to research, staff development, and science education. He received the Distinguished Alumni Award and an honorary doctorate from Bridgewater State University and an honorary doctorate in humane letters from Gratz College in Philadelphia.

Dr. Sousa has been interviewed on the NBC *Today* show, by other television programs, and by National Public Radio about his work with schools using brain research. He makes his home in south Florida.

CHAPTER 1

Introduction

*In an era of parental paranoia, lawsuit mania
and testing frenzy, we are failing to inspire our
children's curiosity, creativity, and imagination.
We are denying them opportunities to tinker,
discover, and explore—in short, to play.*

—Darell Hammond
Philanthropist

IT SHOULD BE SIMPLE. THE MAIN GOAL OF TEACHING IS LEARNING. DURING LEARNING, the human brain makes new neural connections and creates memories for the learner's future use. Learning, by its nature, takes place inside the student's head and is invisible to others. It makes perfect sense that, at some point, it is necessary for teachers to determine through some visible means what learning has occurred. How has the student's brain changed? What knowledge and skills have been processed into long-term memory? This process clearly involves some form of assessment. Ah, and therein lies the rub. How has something so logical and basic to the teaching and learning process as checking what the student has learned morphed into such a contentious issue?

WHAT IS BRAIN-FRIENDLY ASSESSMENT?

This book deals with issues around ways teachers can effectively design and use assessments that are more likely to convince a student's brain that the desired content and skills are worth learning and remembering. A brain-friendly assessment is any assessment device designed and administered by one who has a thorough understanding of how the brain acquires, evaluates, and stores information, as well as the various variables that affect human learning, recall, and performance. A well-designed brain-friendly assessment will reflect a more accurate picture of what the learner knows, understands, and can do than one that is not so designed. For example, the unexpected quiz used as a "gotcha" device will simply reveal how much the learner's cerebral anxiety mechanisms have interfered with long-term memory and recall.

THE CURRENT STATUS OF TESTS AND ASSESSMENTS

Assessment is the most dreaded word in public education. That includes all its synonyms, such as *testing, evaluation, measurement,* and *appraisal.* In recent years, government officials and prominent business leaders have become enamored with test scores. Because they look at schools as a business operation, they insist on strong teacher accountability, use test scores as the primary measure of learning, and encourage competition through charter schools. The reality is that none of these actions will result in better teaching and deeper learning. What they will do is guarantee that teachers will take time away from innovative and interesting student-centered projects and from the nontested subject areas in order to prepare students for the mere basics needed to pass the mandated test. Furthermore, the test scores do not measure the skills and traits—such as creativity, curiosity, higher-order thinking, collaboration, and enthusiasm—that students will need to succeed in college or in a career in this rapidly changing technological world.

ASSESSMENT AND TESTING ARE NOT THE SAME

Assessment and testing are not the same. Although the words are often used interchangeably, we can consider a test as one aspect of assessment. Tests are often of the paper-and-pencil variety, but assessment encompasses other evaluative measures as well. For example, in addition to the written test, a teacher may ask a student to perform a certain task or operation and observe the student's behavior. The teacher may orally ask about the student's thinking process and decision making while the student carries out the task. Students might also submit projects they have designed around a theme they are studying, such as a music composition, a painting, a dance, or a model. The teacher assesses whether the project shows a deep understanding of the concepts under study. Testing, then, is usually an *event,* a snapshot of what the student knows at that moment. But assessment is a *process* whereby the teacher is continually monitoring the student's progress and providing corrective feedback and support as appropriate.

Assessment is a process; testing is an event.

Because high-stakes testing is a one-time event, teachers feel the pressure. An agitated eighth-grade teacher recently remarked, "If my principal reminds me one more time about the importance of the upcoming state-mandated tests, I'll scream!" Such irritation is understandable and widespread. Teachers feel marginalized and profoundly demoralized by this growing movement to publicize their schools' test scores. There is now a backlash developing in school districts. For the past several years, teachers, administrators, and parents have been protesting that standardized tests are unreliable because, among other things, they try to measure in a few hours what a student was supposed to learn over a six- to eight-month period or in several years.

Many standardized tests are administered to students, but high-stakes achievement tests, given by all states except Nebraska, are the most controversial because their scores can result in serious consequences. With low scores, an elementary student may not advance to the next grade, or a high school student may be denied a diploma. There can also be punitive actions against a poor-performing school and its staff, with little due process. In some states, these schools are shut down or turned over to private management. Further, the results may determine the extent of federal and local funding, and which teachers keep their jobs and how much they are paid. To add to the testing pressure, many districts also administer their own midyear and final examinations in the middle school and high school curriculum areas.

FIGURE 1.1   The diagram shows the large number of tests that can be administered to students in US schools. Most of these are standardized tests.

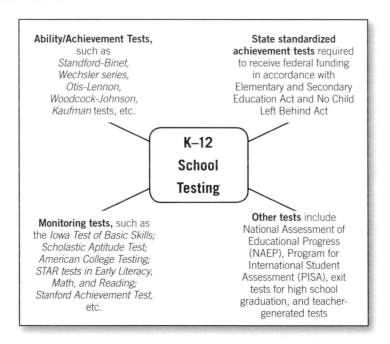

Go to www.learningsciences.com/bookresources to download figures and tables.

In addition to the mandatory tests, many districts participate in testing programs, such as the National Assessment of Educational Progress (NAEP), that compare the students at certain grade levels in the district to those in other schools around the country. International comparisons are made through the Program for International Student Assessment (PISA). Figure 1.1 illustrates the large number of tests that can be administered to different students at various grade levels throughout the academic year.

ATTITUDES TOWARD TESTING

As the pressure for more testing has increased in recent years, attitudes toward testing have become polarized and public. At some point, every segment has spoken out for or against the push for greater teacher, student, and policy-maker accountability. Several studies and numerous articles in the news media have highlighted these opinions and attitudes.

Teacher Attitudes

FIGURE 1.2   This chart shows the attitudes of K–12 teachers when asked if standardized testing has a positive effect on education. They responded on a five-point Likert scale where 1 = Strongly Disagree and 5 = Strongly Agree. *Source:* Schuette et al. (2010).

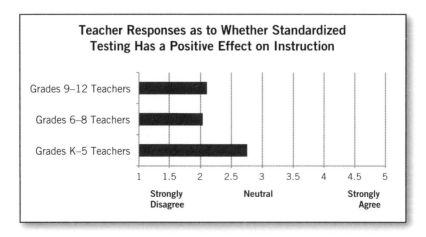

Teacher attitudes toward testing vary among the grade levels and type of instruction (e.g., regular classroom, special education, or gifted). But many teachers generally feel that high-stakes testing has a negative effect on instruction. As an example, one study of nearly two hundred K–12 teachers in public and independent schools showed halfhearted support for standardized tests (Schuette, Wighting, Spaulding, Ponton & Betts 2010). One of the questions they were asked is whether standardized testing has a positive effect on instruction. They responded on a five-point Likert scale where 1 was

Go to www.learningsciences.com/bookresources to download figures and tables.

"Strongly Disagree" and 5 was "Strongly Agree." Figure 1.2 shows the results. All teachers in the study disagreed with this statement to some extent. Elementary school teachers had a slightly less negative attitude toward these tests than did middle school and high school teachers.

Teachers cite high-stakes testing as a major reason for leaving the profession.

One persistently stark statistic is that about 46 percent of teachers leave the field within five years. When these teachers were asked why they left, a majority stated that they had not been properly prepared, had increased demands placed on them because of high-stakes testing, and were not getting adequate support from the school administration in dealing with classroom discipline (Lepi, 2013). This rate of turnover causes a serious disruption in the continuity of instruction and undermines the school's learning climate.

Parent Attitudes

Parents, of course, have been assessing their children since birth. For instance, they keep track of how quickly and how well their children learn language, they measure their changes in height and weight, and monitor the development of their social skills. But parents are now wondering about the value of the school's testing programs. A recent national survey of more than one thousand K–12 parents showed that although many support the *idea* of standardized testing, they question the *process* (Tompson, Benz & Agiesta, 2013). It is interesting to note from this survey that the number of parents who say standardized tests effectively measure their child's performance and the educational quality offered in their schools decreases as the parents' educational attainment level increases.

Despite their support for these tests, parents also say they resent the vast amount of time spent on preparing for and taking the tests as well as the additional stress it places on their children. They get upset when poor test scores result in the replacement of the principal and staff at their neighborhood school. Even more upsetting is when poor test scores result in closing the school and when their children get shipped off to—and have to adjust to—a different school. Parents of elementary school children have also noticed that the sharp focus on the subjects tested, mostly mathematics and reading, have crowded out lessons in science, social studies, and especially the arts. In some high schools, standardized exams can determine a student's course grade in subjects ranging from social studies to mathematics.

*More parents are choosing to exempt
their children from high-stakes testing.*

A survey of recent news reports reveals an unprecedented number of revolts against standardized testing by students, teachers, and parents. In Rhode Island, New York, California, and Oregon, students have used various forms of protests, such as opting out of their state's standardized tests and persuading other students not to take them (Strauss, 2013a). In Seattle, teachers at Garfield High School voted unanimously not to give the district's required reading and mathematics test. They encountered predictable resistance from district officials, including a warning from the district's superintendent. Regardless of the harsh criticism from outside observers, many students and parents sided with the teachers (Shaw, 2013). The student government and the Parent Teacher Student Association voted their support for the teachers' decision. Numerous letters came from parents requesting that the district exempt their children from testing, mainly because they felt the tests were not a fair measure of the effectiveness of their children's teachers.

Administrator Attitudes

Some school principals have also had their fill. More than five hundred principals across New York State have protested the newly revised state assessment, saying that their students are being overtested with little or no clear benefit to their educational achievement (Tyrrell, 2013). The principals' letter to the state department of education noted that state testing has increased dramatically in recent years and that the tests take too long. They claimed that numerous test questions were ambiguous, causing stress among the students. Furthermore, they contend that standardized testing has failed to show how they help children learn despite their high cost to the taxpaying public.

Protests like these have already led to testing rollbacks in Texas, Seattle, and Chicago. In Texas, for example, more than five hundred school boards have passed resolutions requesting the Texas Education Agency to reduce its emphasis on high-stakes standardized testing. Several districts in Florida have done the same. The protests have also prompted some states to pull back their participation in the new assessments required for the Common Core State Standards that have been developed in language arts and mathematics and adopted by a large majority of the states. Concurrently, the US Department of Education is offering some states a two-year delay before they begin a controversial program of new evaluations for teachers using standardized test scores.

Proponents of testing argue that, when put into perspective, all the opposition to high-stakes testing actually represents a small fraction of the concerned parties. They cite studies showing that many students may not be happy about the tests but they recognize the need for them. For example, a 2006 survey of more than 1,300 public school students in grades 6–12 found that 71 percent of the students thought the number of tests they must take is acceptable (Johnson, Arumi & Ott, 2006). Nearly 80 percent said they thought the test questions were fair. Of a similar number of parents who were surveyed, less than 20 percent thought their children were taking too many standardized tests. They add that most major civil rights groups are supporters of rigorous testing. These supporters do admit that school districts have not done an adequate job educating parents about the need for standardized tests to maintain the quality of instruction. Unfortunately, there have been no studies of student and parent attitudes toward standardized testing in recent years. That makes it hard to determine how much the opposition to these tests is growing or what impact it has had. There also have been no major studies that measure how much the testing has increased. Numerous anecdotal reports, however, indicate a widespread increase due to (1) the continued testing requirements of the No Child Left Behind Act (NCLB), (2) pretesting in those states that have already begun implementing the Common Core State Standards, (3) the adoption of state standards and testing into subject areas other than language arts and mathematics, such as science and history, and (4) the increased pressure on states and school districts to use testing results as a measure of teacher accountability.

DO HIGH-STAKES TESTS IMPROVE STUDENT ACHIEVEMENT?

Proponents of high-stakes testing insist that these tests prod teachers into helping students learn what they need to get satisfactory scores. If these scores on high-stakes tests represent true increases in student learning, then one would expect that these same students would demonstrate similar achievement on comparable lower-stakes tests. But the few research studies that have tested this premise show otherwise, and have shown so for at least a decade. For example, an in-depth analysis of test scores in the Chicago Public Schools during the period of 1993 through 2000 showed a significant increase in student achievement scores in mathematics and reading on high-stakes standardized tests (Jacob, 2002). But when these same students were tested using a low-stakes statewide test that measured essentially the same skills, their achievement dropped significantly. The report concluded that the high-stakes results showed little empirical evidence that they can be used to accurately measure student achievement or other forms of accountability. Moreover, the study found that teachers responded to the strong pressure exerted by the upcoming high-states tests by (1) increasing their referrals for placement in special education classes, (2) retaining students in grade before the testing period, and (3) reducing the instructional time for low-stakes subjects, like science and social studies, as indicated by scores in these subjects increasing at a significantly slower rate than those in reading and mathematics.

Research studies do not support the argument that high-stakes testing improves student achievement.

A similar analysis of testing data in twenty-five states also found that the pressure generated for the high-stakes tests for NCLB did not result in higher student achievement scores on the NAEP, which is considered a low-stakes test (Nichols, Glass & Berliner, 2012). The grade four NAEP scores in mathematics did seem to improve after NCLB, but the improvement was at a slower rate than before NCLB. Furthermore, the NAEP scores in reading for students living in poverty declined in those states that attached the highest stakes to the test results.

An extensive review of research studies on all types of testing found that high-stakes tests in general typically measure basic skills and knowledge (Faxon-Mills, Hamilton, Rudnick & Stecher, 2013). Seldom do they ask questions requiring creativity, critical thinking, or imagination. Consequently, following the old adage of "What gets tested gets taught," teachers tailored their instruction to focus on basic knowledge with very little time devoted to higher-order thinking involved in creating projects or analyzing and solving difficult problems. It is no wonder then that this type of rote and shallow learning rarely results in long-term retention. The student's brain quickly becomes accustomed to carrying information in temporary memory long enough to take the test and then forget it. We will discuss more about this undesirable phenomenon later in the book.

Increases in achievement scores on standardized tests do not mean there has been a true acquisition of learning.

It seems that although high-stakes testing may result in an increase in test scores on a particular standardized test, the scores may not indicate that there has been a permanent increase in student learning. In other words, whatever methods the teachers used to produce the results on the high-stakes test—such as enhancing their students' test-taking skills or teaching to the test—did not result in a permanent change in the learner's brain. There was no true acquisition of learning and no long-term memory of it. Is this what we really want the end result of expensive and time-consuming high-stakes tests to be?

WILL CHANGE OCCUR?

Change is in the air, but will it succeed? Are the forces aligned to reduce the amount of high-stakes testing powerful enough to overcome the forces supporting a billion-dollar testing industry? Can the opponents outwit those seeking to ensure that there is robust accountability for the approximately half-trillion dollars spent annually on public education in the United States? Not likely in the short term. Even though it makes sense to argue that less testing allows more time to be devoted to meaningful teaching and learning, and that the current pressure of high-stakes testing is not improving student learning, the reality is that powerful political influences are likely to trump the educational rationale. Like it or not, the current level of testing will probably not change in the foreseeable future. So now the question is whether there are other things educators can do to make the assessment and testing processes more brain friendly and conducive to authentic learning.

If We Are Stuck with It, Let's Improve It

Well-designed, large-scale assessments genuinely contribute data that can be useful in guiding changes in curriculum and instructional strategies. The question is whether their current design provides *reliable* data and whether they need to be so frequent. Current research findings suggest that the answer to both parts of this question is no. Although educators probably cannot do much about the *quantity* of tests being administered, they can at least try to improve their *quality*. Higher-quality tests, of course, are of little value unless they result in improved instruction and higher student achievement. We need, then, to look for ways to create and use assessments that result in observable and meaningful student learning.

One encouraging sign is the announcement from the College Board that the SAT will be significantly redesigned for the spring of 2016. Questions will deal with matter more relevant to today's high school student and more closely tied to what is being taught in the nation's high schools. The test will require students to analyze source materials and cite evidence to justify their answers. Vocabulary words will focus more on words that are widely used in college and careers. This revamp is coming at a time when more colleges are choosing to opt out of using standardized testing scores as a major criterion for admission. Administrators at these colleges say the SAT does not really test what is taught in the nation's high schools, and students never find out which questions they got correct or incorrect, or why, and thus they learn nothing from this frustrating ordeal (Botstein, 2014). Further, they maintain, the SAT scores are becoming far less predictive of who will succeed in college. (Keep in mind that the SAT was designed as an *aptitude* test, not as a measure of student *achievement*.) Despite this trend, the National Association for College Admission Counseling reports that about 80 percent of four-year colleges still rely on these types of scores. What may be changing, however, is how much *weight* the admissions staff gives to these scores when deciding whether they should admit a student.

Don't Forget the Serenity Prayer

Regrettably, the reality is that whatever influence teachers have to redesign their statewide tests (and that is actually occurring in a few states) will take a considerable amount of time—certainly several years. So, with a nondenominational serenity prayer in mind, do not fret over those things you cannot change, and work to improve the things you can change, remembering that you already have the wisdom to know the difference. In accepting this reality, we turn our attention to those assessments over which teachers *do* have considerable influence. These include assessments for which they usually provide input, such as locally generated subject-area midterm and final exams, and those that they create themselves for classroom use. Teachers *can* make these assessments brain friendly and, by doing so, improve the depth of learning, understanding, and skill development needed to be a successful student, in addition to whatever standardized tests purport to measure.

HOW THIS BOOK CAN HELP

This book is *not* a collection of assessment activities. Plenty of books are already available offering that. Rather, this book presents and explains the critical factors that educators should consider when designing and selecting any type of assessment, if the purpose is to accurately disclose what a student has learned. It is intended to empower the educator with the research findings from *educational neuroscience* relating to the conditions under which the human brain is most likely to reveal what it truly knows, comprehends, and can accomplish when asked to do so. Assessments are too often designed and delivered for the convenience of the teacher. Here we suggest how to design and administer assessments that are in the best interest of the student.

Chapter Contents

Chapter 2 addresses several important issues. It starts with *why* we should assess student learning at all. Although the answer may seem obvious, the superficial way in which assessment is often approached belies the intricate activity that the learner must experience when responding to any assessment. Next, it looks at *who* we should assess. Some complicated questions we examine are whether all students should take the same test or whether different tests should be designed according to the students' perceived ability levels. Also, how do we fairly assess those students who are homeschooled? Finally, it tackles the long-standing debate over *what* we should assess. Educators continually argue for teaching students how to do higher-order thinking, how to analyze complex problems and devise potential solutions, and how to be creative. But do standardized and other tests really measure these capabilities? If not, can we design assessments and tests that do?

The answer to that question is discussed in Chapters 3 and 4, which look at *how* we should assess students. Although teachers have little or no control over how national and statewide high-stakes tests are designed and administered, they usually have substantial control over how they design and use their self-selected assessments in their classes. Chapter 5 addresses the questions of *when* and *where* we should assess. These may seem like trivial questions, but are they really? High-stakes tests are given in the early spring in many states. Is this the best timing? As to when we give a test, what are these circadian rhythms we hear about, and do they really matter during test-taking, even if the assessment or test is teacher-generated? We present some interesting research that looks at whether the surroundings affect student performance when taking a test. Finally, Chapter 6 returns to the central theme of how we can design brain-friendly assessments that promote deep understanding and accurately reflect the extent of student learning and retention.

Note about Effect Size

Throughout the book there will be references to *effect sizes* (ES) when describing the degree of influence an intervention or variable has in research studies on student achievement. Effect size not only reveals whether the intervention worked but also how *well* it worked. It is calculated as follows:

$$\text{ES} = \frac{\text{(Mean of experimental group)} - \text{(Mean of control group)}}{\text{Pooled standard deviation of both groups}}$$

For studies in education and the social sciences, effect sizes of 0.25 and greater are usually considered significant, although some researchers, such as Hattie (2012), set a minimum standard of significance at 0.40. The closer the effect size gets to 1.0, the greater the influence of the intervening variable. Effect sizes are particularly useful when comparing the relative sizes of effects from different studies that are researching a common variable—in this case, student achievement.

This will be a surprising journey through the chapters as we dispel a few commonly held beliefs about assessment, so be prepared. My hope is that some of the ideas and evidence presented here will at least provoke educators to rethink what schools are really all about, and what they are meant to be.

CHAPTER 2

Why, Who, and What We Assess

*The test of a good teacher is not how many
questions he can ask his pupils that they will
answer readily, but how many questions he inspires
them to ask him which he finds it hard to answer.*

—Alice Wellington Rollins
American author, 1847–1897

TEACHERS HAVE BEEN ASSESSING THEIR STUDENTS SINCE AT LEAST ANCIENT GREECE. Socrates tested his students through oral questions and conversations. Their answers, whether right or wrong, led to more dialogue, insights, and greater depth of understanding. No one worried about getting a score. Even today, educators profess that the Socratic approach represents the epitome of excellence in teaching and learning. Yet, look how far we have strayed from the brain-friendly and challenging Socratic method to the labyrinth of standardized and other testing that has become today's norm.

MAJOR PURPOSES OF ASSESSMENT

In an ideal world, the major reason teachers use assessments would be to determine how much a student has learned after a period of instruction. Any other reason would be a distant second. But, alas, in this world of education that's obsessed with scores, one wonders if assessing learning is anywhere near first place. Administrators can use assessments and test scores for placement of students, and policy makers and parents can use them to compare schools within a jurisdiction and appraise teacher accountability. But the purpose of this book is to focus on the most important reason for assessment: to determine how much and how well students are learning as they progress through our schools. I suggest that there are four major components of student-centered, brain-friendly assessments. As Figure 2.1 illustrates, in addition to assessing student growth, we assess to determine teacher effectiveness, to ensure that we are matching instruction with curriculum, and to examine the school's climate to entice students to stay and not drop out.

Student Growth

Obviously, students and teachers are central to the teaching–learning process. Assessment for instructional purposes should focus on these two groups. Essentially, we are looking to see how well students have progressed toward learning goals and how effective teachers have been in fostering that progress.

FIGURE 2.1   The diagram illustrates the major purposes of assessment.

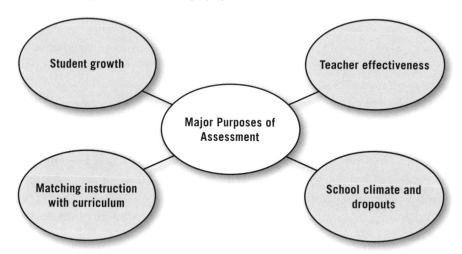

Assessment is a brain-centered activity. For years, I have been describing teachers as brain changers. Their job is essentially to present information and demonstrate skills in such a way that a learner acquires that information and skill. For this to happen, the learner's brain must change. It must create or rewire neural connections that commit the information and skill to long-term memory and have it available for recall. It usually takes several tries for this to happen, and that is why we have students practice what they have just learned. *Practice* helps strengthen the new neural connections and, if possible, consolidate the new information with what the student has already learned. Assessment, then, provides the learner an opportunity to demonstrate understanding of the information and mastery of the skill.

Teachers are brain changers!

Go to www.learningsciences.com/bookresources to download figures and tables.

The results of these assessments show each student's progress toward short- and long-term curriculum goals. They help teachers diagnose an individual student's strengths and weaknesses, and respond accordingly. This is the essence of differentiated instruction that can be effective without being burdensome (Sousa & Tomlinson, 2011). Teachers can also use assessment results to give all-important feedback to students about their progress, areas that need more work, and what they have learned well. Feedback can increase students' *motivation* by serving as an indicator of their progress. We will discuss more about the power and types of feedback in Chapter 3.

> *Nonthreatening assessments give a teacher insight into how quickly and how well an individual student learns.*

When students see assessments as learning activities rather than "gotcha" devices, they tend to be better motivated to learn and to set goals for themselves. Well-designed assessments help students practice what they know. This practice leads to reinforcement and retention of learning. The more often a learner recalls information and practices a skill, the more likely it is to be consolidated firmly into long-term memory and, even more importantly, associated into networks of past learnings and skills. Furthermore, keep in mind that a teacher cannot motivate a student whom the teacher does not really know. Frequent, nonthreatening assessments give a teacher insight into how quickly and how well an individual student learns.

Teacher Effectiveness

Student assessments also provide feedback to teachers by indicating the effectiveness of their instruction. If a majority of students did poorly on a particular concept, it can indicate that the instruction was incomplete, too rapid, or inadequate or the topic was too difficult. Perhaps the teacher did little or no checking for understanding before moving on to the next topic. If nearly every student got high marks on an assessment, it could indicate first-rate instructional skills, high student performance, or that the topic's level of difficulty was too easy. Teachers should explore and analyze these possibilities in order to decide on future instructional strategies. The results of student assessments also tell us a lot about curriculum. Analyzing how well or how poorly the students fared on test items may indicate the appropriateness and level of difficulty of curriculum topics. Item analysis can identify content that may no longer be relevant or be too complex or too simple for the class's age group. Assessments help teachers adjust the focus and pace of instruction and learning.

Matching Instruction with Curriculum

Assessment results allow educators to determine how closely a match exists between what teachers are supposed to be teaching (curriculum) and what students appear to be learning. This is not an easy task, because there are so many variables in the teaching and learning process that can affect teacher and student performance. The results may also suggest that some new instructional strategies should be considered, or current ones abandoned, to improve the curriculum-learning connection. With so many "how-to" books containing instructional activities available, it is difficult to know which ones are brain friendly and effective. This is where action research comes in. Teachers can try out a strategy in a systematic way to determine its effectiveness in the classroom.

When assessments include open-ended questions that allow students to explain the depth of what they know about a topic, educators can get clues about what curriculum changes may need to be made to propel students to high-order thinking, analysis, and real-world problem solving. A student's explanation can sometimes provide astonishing insights into a topic that would benefit all learners.

School Climate and Dropouts

Not only are assessments helpful in revealing what students know, understand, and can do, but also they are useful in determining how students—especially those who drop out of school—*feel* about their school experience. Do they feel welcomed or excluded? Does it seem that teachers really care about their success or are they indifferent? Are students bullied? Is school interesting or boring? Is their work challenging? This information is important to collect because the nation's dropout rate in 2013 hovered around 8 percent for all students, but double that for Hispanic students. In 2013, more than 3 million students dropped out of high school. That is about 8,300 per day. Although the dropout rate has been slowly declining in recent years, there are still too many students, particularly black and Hispanic students, out looking for work with no high school diploma (see Figure 2.2).

Of particular interest is *why* students drop out of school. In a survey supported by the Gates Foundation, nearly five hundred high school dropouts gave the following five reasons as major factors for leaving school (Bridgeland, DiIulio & Morison, 2006):

- 47 percent: Classes were not interesting

- 43 percent: Missed too many days to catch up

- 42 percent: Spent time with people who were not interested in school

- • 38 percent: Had too much freedom and not enough rules in my life

- • 35 percent: Was failing in school

FIGURE 2.2   The graph shows the percent of students who dropped out of high school in years 2005 and 2013. The trend is downward, but the number is still too high. *Source:* Statistic Brain (2014).

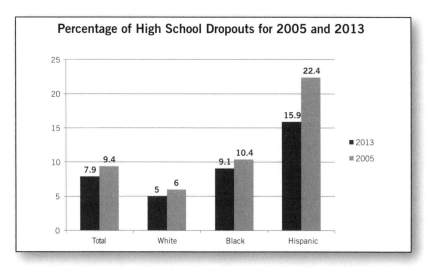

The results of school climate and school attitude assessments might have revealed how many students do not find school interesting or relevant. This could lead to appropriate curricular and instructional changes and may prevent potential dropouts from walking out the door.

FIGURE 2.3   The chart shows the percentage of recent high school graduates who responded to the question, How well did your high school education prepare you to be successful in your first full-time job? Adapted from Horn et al. (2012).

Go to www.learningsciences.com/bookresources to download figures and tables.

Regrettably, it seems that high schools are not adequately preparing their graduates for a postsecondary career or for college. A national study conducted by Rutgers University asked 544 high school graduates how well they felt their high school had prepared them to be *successful* in their first full-time job. Figure 2.3 shows their responses. Only 42 percent thought that the high school had prepared them "extremely well" or "pretty well" to be successful in their first full-time job (Horn, Zukin, Szeltner & Stone, 2012). Another 42 percent felt their high school prepared them "not very well" or "not well at all." One can interpret these results in several ways. They could suggest that the school's curriculum did not include the knowledge and skills that students needed to obtain and keep a job in this highly technological world. Another possibility is that, although the students did graduate from high school, they may not have had the motivation to benefit from their courses by discovering how to apply what they learned in a workplace environment. A third interpretation is that the instructional level in the courses was too basic and did not present students with options to apply their knowledge and skills to solve problems and think critically. Appropriate assessments might have provided clues to one or all of these possibilities.

A report released by ACT, a leading national testing company, showed that only 25 percent of the 1.67 million students who took the comprehensive ACT test in 2012 met all of the college-readiness benchmarks in English, reading, mathematics, and science. According to ACT, this number represents about 52 percent of all 2012 high school graduates. Here again, multiple explanations are possible. Perhaps the standardized test is not measuring what is taught in today's schools. Is the passing score realistic? Are high school standards too low? Is the instruction adequate? Any or all of these possibilities need an explanation. Regrettably, the test results just tell *what* happened, not *why*. So, teachers should design their own assessments to reveal why students performed as they did so they can use that information to make any instructional adjustments needed to improve student achievement. But to do this accurately and fairly, the assessments must be carefully designed to be brain friendly, so students will not fear them but rather perceive the assessments as indicators of their academic progress.

MATCHING ASSESSMENTS TO STUDENT ABILITIES

Students come to a school with varying levels of ability and motivation, with different interests, from numerous cultures, and speaking a wide variety of languages. Teaching this marvelous medley of individuals is indeed a challenge but assessing them may be even more so. Given the mix of cultures, languages, abilities, and interests, one may pose the question, should all students be assessed? The answer simply is yes, but not necessarily in the same way. The brains of all these students are wired differently as a result of their experiences. Even identical twins raised in the same environment will have significant differences in how their neural networks are connected and how they interact with their world, though they look and perhaps behave so much alike.

Artistically talented students should be able to demonstrate their learning through the arts.

It has taken educational leaders and other stakeholders a long time to realize these differences and that one size does not fit all. Equal opportunity is different from equal treatment. Gifted students should be pushed to meet their intellectual potential rather than wait for others who need more time to learn the same material or skills. Struggling students should be given extra help to succeed in their work rather than facing intimidating assessments or tests they cannot pass. Artistically talented students should have opportunities to demonstrate their learning and creativity through the arts. Unfortunately, in many schools, the classes in the arts are being reduced or eliminated to prepare for high-stakes testing or to devote more time for science and mathematics instruction as part of the STEM (science, technology, engineering, and mathematics) initiative. This trend is particularly troubling because recent brain studies suggest that creativity can be taught and developed. All the arts are forms of creative expression, and practice in them serves to develop attention, spatial skills, working memory, and persistence (Sousa & Pilecki, 2013).

Students with Special Needs

Students who have special needs still have a right to an educational experience that will help them reach their full potential. They also have a right to know how well they are progressing. But in some states, the testing process for these students is bordering on the absurd. In upstate New York, a fourth-grade boy undergoing prescreening for brain surgery and hooked up to an *electroencephalograph* (EEG) was asked to take a state test from his hospital bed (Strauss, 2013b). Despite his recurring seizures from life-threatening epilepsy, a teacher was dispatched to the hospital to administer the fourth-grade state test. Meanwhile, a nine-year-old Florida boy, who was born with just a brain stem and cannot speak or see, was required to take the state test. A state employee read it to him as though the boy could actually understand it (Rowland, 2013).

Some accommodations may need to be made when assessing or testing the learning of students with special needs. Curiously, some people balk at making such accommodations, arguing that standardized tests should be administered in a standardized way for all students. Doing so, however, means that students with certain disabilities will have test scores that may be deceptively low. Both teachers and students become frustrated when students make genuine progress in their academic achievement but are unable to demonstrate that progress on a written test. The purpose of an accommodation is not to increase the scores of students with special needs, but to provide an environment that will help them attain a score that accurately reflects their true understanding and proficiency.

English Language Learners

Carefully assessing English language learners is important because these students are faced with a double challenge. Not only must they acquire the information and skills needed for their development and academic achievement, but also they must do so through a language that is not their own. Until their brains have mastered a sufficient amount of English, they are constantly using up neural energy to translate back and forth between English and their native language. This is a laborious process that some English language learners find frustrating. Assessment of these students requires measuring their English language proficiency as well as their academic progress.

WHAT SHOULD WE ASSESS?

We now turn our focus to what brain-friendly, teacher-prepared assessments should measure. Officially, they should reveal the ongoing progress that students are making toward identified learning objectives and standards. But because learning is a brain-centered phenomenon, when teachers give any type of assessment of learning, what are they measuring? In essence, the assessments measure how much of the acquired learning actually was processed by the brain and how much was consolidated and retained in the student's long-term memory. To understand that process, let's take a look at how we believe memory systems work.

Assessments should be determining how much of the learning has been processed into long-term memory.

Memory Systems

How best to classify forms of memory is a source of continual debate, especially as the increase in brain research studies offer up new findings and insights. Case studies of people experiencing memory loss, experiments designed to test memory, and analysis of brain scans all suggest that memories exist in different forms. The problem is getting neuroscientists to agree on a fixed taxonomy that defines the stages and types of human memory. And, of course, as new research results emerge, the taxonomy and nomenclature change accordingly. Figure 2.4 represents what most active researchers seem to accept as a workable model for describing memory systems at the time of publication.

FIGURE 2.4   The diagram is a representation of how memory systems are related.

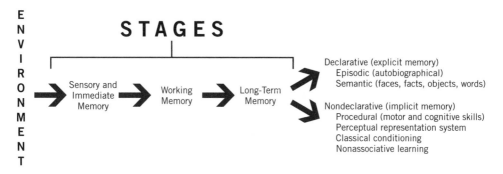

Stages of Memory

It is important to keep in mind that memory is a very complex phenomenon involving many different areas of the brain at various levels of processing. We look first at stages of memory, which are conveniently defined based on the amount of *time* that information is processed. The three stages of memory are the following: sensory/immediate, working, and long-term.

Sensory Memory

Sensory and immediate memory share many common characteristics and are often treated as one entity. However, there are some slight differences in how they respond to external stimuli. *Sensory memory* monitors the nature and strength of incoming sensory impulses for survival content. In just milliseconds (one one-thousandth of a second), it uses the individual's past experiences to determine the degree of threat or importance. Most of the sensory signals are unimportant, so they are dropped out of the memory system. For example, have you ever noticed how you can be studying in a room

Go to www.learningsciences.com/bookresources to download figures and tables.

where there is construction noise outside? Eventually, you no longer hear the noise because the sensory register is blocking your consciousness from paying attention to this stimulus. On the other hand, if you hear the fire alarm sounding, the sensory register instantly alerts the brain to the threat, and you refocus to determine your course of action.

Immediate Memory

Immediate memory is a convenient place to hold information that you need for just a few seconds and is not apt to be saved. For example, when you look up the telephone number of the local pizza place for a delivery, you usually can remember the number just long enough to make the call. When the call is finished, you no longer need the number, and it drops out of the memory system. The next time you call that pizza place, you will most likely have to look up the number again. Why is that? Very simply, you cannot recall information that your brain does not retain. However, if you keep getting a busy signal when calling the pizza place, then after a number of tries the information moves into working memory where it can be retained for a longer period of time.

You cannot recall information
that your brain does not retain.

Working Memory

Working memory is the busy place where conscious processing occurs. Working memory is the area of limited capacity where we can build, take apart, or rework ideas for eventual storage somewhere else. When something is in working memory, it generally captures our focus and demands our attention. Information in working memory can come from the sensory/ immediate memories or be retrieved from long-term memory. Brain imaging studies show that most of working memory's activity occurs in the *frontal lobes* (just behind the forehead), although other parts of the brain are often called into action.

The functional capacity of working memory
now seems less than previously thought.

Studies in the 1950s through the 1970s seem to indicate that working memory had a functional capacity limit that varied with age. Young children and preadolescents were thought to have an average capacity limit of about five items, plus or minus two. In adolescents and adults, the average capacity increased to about seven items, also plus or minus two. These capacity limits have been the standard that most cognitive psychologists have used. More recent research, however, has raised questions about the exact capacity limit of working memory. Some studies suggest that it is less than previously thought and may now be three to five items for adults. A few others say it is difficult to state an actual number because variables such as interest, mental time delays, and distractions may undermine and invalidate experimental attempts to find a capacity limit (Cowan, 2010). Nonetheless, most of the research evidence to date supports the notion that working memory has a functional limit and that the actual number varies with the learner's age and the type of input (factual information, visual, auditory, etc.) and the nature of the cognitive processing (Price, Colflesh, Cerella & Verhaeghen, 2014).

Can you see the implication this functional capacity has on lesson planning and assessment? Obviously, the elementary teacher who expects students to remember in one lesson the eight rules for using the comma is already in trouble. So is the high school or college teacher who wants students to learn in one lesson the names and locations of the ten most important rivers in the world. Keeping the number of items in a lesson objective within the appropriate capacity of working memory increases the likelihood that students will remember more of what they learned. Less is more! It also means that assessments or tests that require students to recall large amounts of information simultaneously to find an answer are not brain friendly, because they may overload working memory's capacity. The student's mental frustration with this overload situation results in an answer that is not an accurate measure of what the student really has learned.

Long-Term Memory

Some stimuli that are processed in the temporary memories are eventually transferred and encoded into long-term memory sites. In doing so, they actually change the structure of the brain cells (*neurons*) so the memories can last up to a lifetime. This encoding process occurs during sleep—specifically during the rapid eye movement (REM) cycles. Figure 2.5 illustrates the average REM periods during a normal eight-hour sleep cycle, although the REM periods can vary widely in some individuals. It should be noted from the chart that getting only five hours of sleep means that the average person loses at least one full REM cycle and perhaps part of another. Therefore, the brain has less time to encode information from working memory into long-term memory, and that unencoded information may be lost upon awakening—a good reason to get a full night's sleep!

FIGURE 2.5   The hashed boxes on the graph show the REM periods during an eight-hour sleep cycle. Research studies suggest that information is encoded into long-term memory during these periods. The numbers on the left indicate the stages of sleep from light sleep (1) to deep sleep (4).

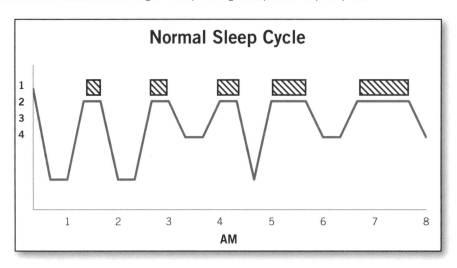

Although neuroscientists are not in total agreement with psychologists as to all of the characteristics of long-term memory, there is considerable agreement on some of their types, and their description is important to understand before setting out to design learning activities and brain-friendly assessments.

Types of Memory

Long-term memory is an intentionally vague term because "long-term" can mean different things to different people. Scientists used to believe that almost anything that was encoded into long-term memory was there for a person's lifetime. Newer research, however, has shown that long-term memories normally weaken over time due to lack of recall, undesirable deposits in brain cells that block neural connections, and the gradual deterioration of white matter that links different parts of the brain. Nonetheless, memories can certainly last a long time. Long-term memory is divided into two major types: declarative memory and nondeclarative memory.

Declarative Memory

Declarative memory (also called conscious or explicit memory) describes the remembering of names, facts, music, and objects (e.g., where you live and the kind of car you own). Think for a moment about a person who is now important in your life. Try to recall that person's image, voice, and mannerisms. Then think of an important event you both attended, one with an emotional connection, such as a concert, wedding, or funeral. Once you have the context in mind, note how easily other components of the memory come together.

Go to www.learningsciences.com/bookresources to download figures and tables.

This is declarative memory in its most common form—a conscious and almost effortless recall. Declarative memory can be further divided into episodic memory and semantic memory.

1.   *Episodic memory* is the conscious memory of events in our own life history, such as our sixteenth birthday party, falling off a bicycle, or what we had for dinner last evening. It helps us identify the time and place when an event happened, and it gives us a sense of self. Episodic memory is the memory of remembering.

2.   *Semantic memory* is knowledge of facts and data that may not be related to any event. It is knowing that the Eiffel Tower is in Paris, how to tell time, how to multiply two numbers, and who the fortieth president was. Semantic memory is the memory of knowing. A veteran knowing that there was a Vietnam War in the 1970s is using semantic memory; remembering his experiences in that war is episodic memory. This is the memory that a student primarily accesses when taking an assessment or test with multiple-choice, single-answer, and fill-in-the-blank questions.

Nondeclarative Memory

Nondeclarative memory (sometimes called implicit memory) describes all memories that are not declarative memories—that is, memories that can be used for things that cannot be declared or explained in any straightforward manner. For example, you use declarative memory to recall your Social Security number (a set of numbers in a certain order), but nondeclarative memory to remember how to ride a bicycle. To date, there are four generally accepted categories of nondeclarative memory:

1.   *Procedural memory:* Refers to the learning of motor and cognitive skills and remembering how to do something, like riding a bicycle, driving a car, swinging a tennis racket, and tying a shoelace. With practice, these skills become almost automatic and require little conscious thought.

2.   *Perceptual representation system (PRS):* Refers to the structure and form of words and objects in memory that can be prompted by prior experience, including our ability to complete fragments of words or tell whether objects in drawings could exist in the real world. Being able to complete this phrase— "A_l's w_l_ _ha_ e_ds _e_l"—and recognize that the object on the right could not exist in the real world are examples of your PRS in action.

3. *Classical conditioning:* Also called Pavlovian conditioning, this occurs when a conditioned stimulus prompts an unconditioned response. Because the response gets associated with the stimulus, this form of learning is called associative learning. Experienced teachers know exactly how to respond in school when the fire alarm sounds. They have learned to associate the sound of the fire alarm with the procedures needed to safely evacuate the building.

4. *Nonassociative learning:* This type helps us to learn not to respond to things that don't require conscious attention, and accustom ourselves to the clothes we wear, the daily noisy traffic outside the school, or a ticking clock in the den. This adjustment to the environment allows the brain to screen out unimportant stimuli so it can focus on those that matter. In sensitization, we increase our response to a particularly noxious or threatening stimulus. For example, Californians who have been through an earthquake tend to respond quickly and vigorously to any weak noise or vibration thereafter, even though it may be unrelated to an earthquake.

Now that we understand the current views about how scientists believe memory works, we can address what we are assessing in those cerebral systems. We already noted that assessment is more than just testing. So what we assess needs to be more than just the information that students acquire in the classroom. We should assess as many aspects of the school experience as we can to ensure a safe, effective, productive, and successful learning environment. So let's look at what we should assess in the brains of the students.

Beware the Working Memory Gambit

Nearly every experienced teacher has lived through the following scenario. The teacher presents a unit of instruction and works with students over several weeks to complete the unit's objectives. After giving a written test on the unit, the teacher finds that nearly all students achieved the unit's learning objectives successfully. With the glow of that satisfying result, the teacher moves on. Several months later, the teacher asks the same class, "Do you remember what you learned in that unit we studied several months ago?" The teacher is met with blank stares and comments such as "What unit?" or "We didn't have that unit!" Stricken with disbelief, the teacher is speechless. What happened here, assuming the students are being honest?

One of the discoveries made in recent years about working memory is that it can hold for a period of time information that may never move to long-term memory. When the information is no longer useful, it drops out of the working memory system. Figure 2.6 illustrates what may have happened in the previous scenario. The students loaded information that they believed was going to be on the test into their working memories, most likely the night before. After completing the test, they assumed the information was no longer needed, and it faded from working memory. In other words, the learning objectives were never encoded into long-term memory, which is why the students reacted as they did several months later. As we noted earlier, our brain cannot later recall what was not stored.

FIGURE 2.6   Information can remain in working memory until it is no longer needed, and then be dropped out of the system. It may never be encoded into long-term memory.

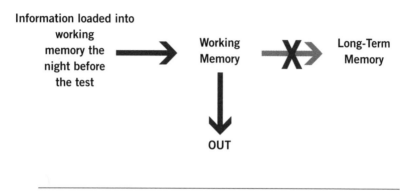

The results of a unit test can be very misleading.

As is frequently the case, the unit test that the teacher gave in this scenario did not reveal reality. Test results were misleading because they implied that most of the students had achieved and remembered the unit's learning objectives and that they were able to recall them during the test. In fact, much of what the teacher thought the students learned and remembered was most likely just placed in their written or electronic notebooks for referral just before the test. What does a teacher do to prevent this scene from being repeated many times?

Criteria for Long-Term Storage

Whenever teachers are asked how long they would like their students to remember what they taught them, the answer is universally the same: "Forever." Yet teachers know that does not happen, as evidenced by the oft-repeated scenario described previously. Whenever a learning episode is coming to a close, the learner's brain has a decision to make: either tag the new information for storage or drop it out of the system. How does it make that decision? It seems that the decision is based largely on two criteria: *sense* and *meaning*.

The brain is more likely to store information if it makes sense and has meaning.

Making sense indicates that the student understands the new information, based mainly on past experiences. It tends to fit into what the learner knows about how the world works. Meaning refers to whether the information has relevance to the student. The student may think, "I understand this, but what does it have to do with me?" A teacher knows that the students have not found meaning when they ask, "Why do I have to know this?" or "When will I ever use this?" Teachers generally focus on presenting the information so that it makes sense but do not give sufficient time to establish meaning.

If the student's brain perceives that the new information does not make sense or have meaning, the chances of it getting stored are low. If either sense *or* meaning is present, the chances increase. If both sense *and* meaning are present, the likelihood of the new information getting encoded into long-term memory is very high (see Figure 2.7). When long-term storage occurs, the students are much more apt to recall the unit they studied several months earlier than say they do not remember it.

FIGURE 2.7   Information is likely to be encoded from working memory to long-term memory if it makes sense and has meaning.

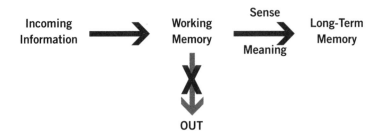

Brain-friendly assessments of students should focus on determining how their brains have filtered, consolidated, and connected the new learnings into their cerebral networks. We should not be measuring just content acquisition. Rather, we should be discovering the ways students can process and manipulate their knowledge and skills to deal with new problems and issues associated with what they have learned. To do this, we need to examine the ways in

Go to www.learningsciences.com/bookresources to download figures and tables.

which the brain can process learning. During the last century, cognitive psychologists have developed a number of models to illustrate how the brain processes learning. Many of them have common features and overlapping descriptions. I still maintain that the classic model developed by Benjamin Bloom, and revised to reflect current research, is a simple, user-friendly representation of human thought processes.

Bloom's Revised Taxonomy

Bloom's original model from the 1950s consisted of six levels of complexity of cognition. From least to most complex, they were knowledge, comprehension, application, analysis, synthesis, and evaluation (Bloom, Engelhart, Furst, Hill & Krathwohl, 1956). From 1995 through 2000, a group of educators worked to revise the original taxonomy based on more recent understandings about cerebral processing. The group published the results of their work in 2001, and the basic revision is shown in Figure 2.8, along with some terms that describe the kind of processing that occurs at each level (Anderson et al., 2001). Note that the levels are in a dotted outline, suggesting that an individual may move among the levels during extensive processing. For example, suppose you ask students to consider the question, "Specifically, why do you think capital punishment does or does not deter crime?" To answer this, they would need to access the Remember and Understand levels to get the definition of capital punishment and comprehend its meaning. They would also do some Analysis and Evaluation to process the pros and cons of capital punishment. They may collect additional data on whether the crime rate is higher or lower in states with capital punishment versus the rate in those that do not have it. Here the students are moving among various levels of complexity while generating their answers.

FIGURE 2.8 Bloom's revised taxonomy places creativity at the top. Next to each level are some terms that describe the type of processing occurring at that level. The dotted outline suggests that an individual may move among the levels during extensive processing. *Source:* Anderson et al. (2001).

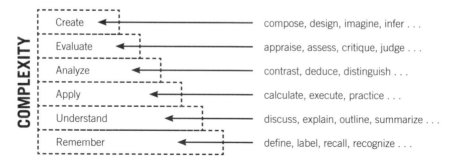

Go to www.learningsciences.com/bookresources to download figures and tables.

Beyond Bloom's Taxonomy

Bloom's taxonomy is a useful tool in understanding the complexity of human thought, but it is not all-inclusive. There are some other levels of complex thinking that cognitive researchers say go beyond the upper limits of this taxonomy. One such level is metacognition.

Metacognition

Metacognition is frequently defined simply as "thinking about thinking." But metacognition is more complex than that. People often do reflective thinking; that is, they *reflect back* on how they solved a problem and decide if there was a better way. Hospital physicians gather after working with a patient to determine if there would have been a better course of treatment. Teachers think back to a lesson that may not have gone as expected and wonder what should be changed next time. Reflective thinking is *after* the fact. Metacognition refers to higher-order thinking where an individual has conscious control over the cognitive processes *while* learning. It is *during* the fact, not *after* it. Students use metacognition when they realize they are having more difficulty learning one concept over another as they are learning it. They also use it when they think about their own thinking processes, such as the study skills they use, their memory capabilities, and how well they monitor their learning as they complete a task.

There are different types of metacognitive knowledge. One type is *person knowledge* (declarative knowledge and memory), which refers to understanding one's own capabilities. *Task knowledge* (procedural knowledge and memory) is another type, and it refers to how one perceives the difficulty of a specific learning task. A third type is *strategic knowledge* (conditional knowledge), which refers to one's own capability for using strategies to learn new information and skills.

Automaticity

A person who has mastered a skill can often perform it without any conscious deliberation, as though it was automated. This form of metacognition is referred to as *automaticity*. For example, a masterful basketball player dribbling the ball down the court is constantly checking the location of each of his teammates and opponents, monitoring and altering his path accordingly, all while eyeing the basket. These split-second decisions show metacognition at work: little conscious thought but masterful performance. Similarly, expert teachers are constantly self-monitoring in the classroom, watching how well their presentation is being received, moving to strategic areas of the room, keeping an eye on the off-task student, remembering time limitations, and making appropriate mini-changes in instruction as the lesson progresses.

Automaticity may appear like a lower-order process, but it is a complement to higher-order thinking. Working together, they help a learner accomplish complex tasks successfully. For instance, reading requires decoding skills, which become automatic with practice. But higher-order thinking is also needed to comprehend and analyze the content and make connections to other concepts in the reader's neural networks.

Brain scanning images suggest that metacognitive monitoring and control are functions of the brain's prefrontal cortex (behind the forehead), which receives sensory signals from other cerebral regions and exerts control (Schwartz & Bacon, 2008; Shimamura, 2008). Young children's brains are not particularly good at this, because their prefrontal cortex is still immature. It is not until fifth or sixth grade that students have sufficient frontal lobe development to begin to develop the understanding of learning strategies that will be effective for them. Studies of older students have found that students who learn about and use metacognitive strategies show greater academic achievement than students who do not (e.g., Sperling, Howard, Staley & DuBois, 2004; Young & Fry, 2008). Because of this positive impact on achievement, students should learn metacognitive skills in both middle and high schools, and teachers should assess these skills on a regular basis.

With Bloom's taxonomy in mind along with metacognition, let us look at what components of learning we are measuring. Essentially, we are assessing three areas: achievement, performance, and proficiency.

Assessing for Achievement

Achievement assessments measure how much the students know and understand about a particular unit of instruction. Can they recall and explain important information about the unit's content? Do they understand the unit's concepts? Can they give examples? Can they connect the new learning to what they already know and can do? Have they mastered the learning objectives? If this assessment is essentially a multiple-choice or one-answer test, then the student is mostly at Bloom's Remember and Understand levels when responding. The type of assessment the teacher constructs will reveal what the student has learned at a particular level of complexity.

Assessing for Performance

Performance assessments, also called authentic assessments, measure the students' ability to perform actual or simulated real-world tasks using the knowledge and skills they learned during one or several units of instruction. It asks the basic question, can the students successfully use what they know and understand? With this type of assessment, students are processing their learning at Bloom's Apply level and perhaps higher. Although students can perform these assessments alone, they often involve students interacting with their peers.

These assessments give students the opportunity to demonstrate what they have learned and apply that learning to new situations. This performance option is particularly beneficial for students who have difficulty writing what they know in a typical paper-and-pencil test. Performance assessments also tend to be closely related to real-world problems and are therefore more engaging and meaningful for students. If student motivation to perform the task is high, then achievement is likely to be better as well.

Assessing for Proficiency

Proficiency assessments do not measure learning acquired in a specific unit. Rather, they measure the students' overall ability to perform in a real-world situation and an unrehearsed setting. They are often used to assess a student's progress toward learning another language. For example: How extensive is the student's new language vocabulary? Does the student use words accurately and with correct syntax? Is the student's proficiency in academic language adequate to understand and explain curriculum content?

There are other valuable uses of proficiency assessments. They may ask a science student to design, set up, and carry out a laboratory experiment that the student was not anticipating. This may seem like just a performance assessment, but it is more than that. Surely, it includes a measure of performance, but it is broader in that it requires the student to think back through several units or even courses to successfully complete the task. The student might be very proficient at designing the experiment and gathering the materials but may not perform well in carrying out the actual experiment.

THE INFLUENCE OF MINDSETS

One of the factors that seem to have a significant impact on student achievement is *mindset*. Mindsets are the beliefs, assumptions, and expectations we possess that direct how we view ourselves and how we interact with others. These mindsets develop when we are very young and are influenced by interactions with our parents, friends, and elements of our particular culture. We store summaries of these interactions in our brains, and our cognitive processing system constantly reviews these summaries to determine how we should respond in future interactions. Over time, these summaries become so consolidated in our memory system that we react almost reflexively when faced with similar situations.

We develop mindsets about many things, such as politics, religion, our family members, our futures, and just about anybody we interact with regularly. Mindsets in adults are so well established in neural networks that they become very difficult to change. Research in neuroscience has found that the neural networking of mindsets is very complex (Mitchell, Banaji & Macrae, 2005). It takes much more neural effort to change one part of a mindset network than to change the entire network (Diamond, 2009). These findings imply that high motivation and considerable persistence are needed to change a mindset, but it can be done. One of the leading researchers on mindsets about learning is psychologist Carol Dweck. She has been studying particularly what it means to be smart and how success happens. Her findings are very important for educators because they remind us how our beliefs and preconceptions about how we learn affect our achievement in school and life.

Types of Mindsets

Dweck's research suggests that we develop at a young age either a fixed or growth mindset about our ability and success in what we do (Dweck, 2006). See Figure 2.9. Individuals with a fixed mindset believe that their success comes from ability and that either you have the ability in a certain domain or you do not. If you do have it, you are all set; if not, well, so be it. Although the environment (nurture) can contribute to your prospects for success, genetic predispositions (nature) pretty much determine how successful you will be in what you pursue. Students with fixed mindsets often say, "I can't do math," "No one in my family was any good at math in school," or "I have no talent at all for music." They attribute their lack of success to their genes, and those cannot be changed. Because they have a desire to look smart, students with a fixed mindset avoid challenges for fear of failing, which has an impact on their self-image. Similarly, they give up easily when faced with obstacles because they believe that effort does not pay off. Consequently, if they retain a fixed mindset, these students are very likely to achieve less than their full potential in school and life.

Students with a growth mindset, on the other hand, believe that although genetics may establish a starting point for a particular ability, their own efforts and persistence are what really determine their success. They rely more on nurture than nature. This viewpoint leads to a desire to improve and creates an appreciation for learning and a resilience that is necessary for achievement. They embrace challenges and persist when faced with setbacks because their self-image is not tied to what others think of them. Further, they see effort not as useless but as the path to mastery. Consequently, these students will continue to reach higher levels of achievement.

FIGURE 2.9   The chart shows the attributes of fixed and growth mindsets.

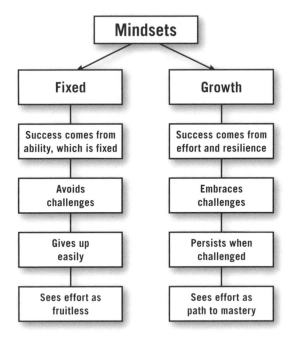

Should We Assess Mindsets?

The good news is that it is possible for students to change from a fixed to a growth mindset. Fortunately, the younger they are, the easier it is to change their mindsets. Therefore, it would be helpful for teachers to know the mindsets of their students, especially those that are fixed. Teachers need to encourage these students and insist on student effort and growth. Over time, as students see evidence that effort leads to success, each student's mindset can change from fixed to growth. Suggestions on how to assess students' mindsets are found in Chapter 3.

Now comes the tough part. Exactly how do we assess students in a brain-friendly way and still meet the various requirements that policy makers and others are demanding? In the next two chapters, we tackle the question of how to assess students.

Go to www.learningsciences.com/bookresources to download figures and tables.

Designing and Using Preassessments and Formative Assessments

I don't think there's any way to build a multiple-choice question that allows students to show what they can do with what they know.

—Roger Farr
Emeritus Professor of Education
Indiana University

HOW TO ASSESS LEARNING HAS BEEN A HOT TOPIC OF DISCUSSION FOR DECADES AND even more so in recent years, given the national focus on school and teacher accountability and the emphasis on more high-stakes testing. We already noted in earlier chapters that the current standardized and teacher-made tests might not be accurately measuring whether the students actually encoded the information and skills into long-term memory. Further, even if the brain did retain these learnings, there is little evidence that students are able to apply them to real-world situations or use them to succeed in college or career.

Recall that we are focusing here on teacher-prepared and teacher-selected assessments. It is interesting to note that the root of the word *assessment* comes from the Latin *assidere*, which means "to sit beside." Ideally, the teacher should sit beside each student, perhaps on a stone bench as Socrates did, and ask appropriate questions to determine how much the student has learned. In the days of Socrates, and later in the one-room schoolhouse, this was very possible. However, with class sizes ranging from twenty-five to forty students or more, time limits require that teachers use other assessment methods. Nonetheless, they can still be brain friendly.

The influence of past learning on new learning is called transfer, and it is a powerful force.

Brain-friendly assessments measure the progress a student's neural networks are making toward acquiring and understanding new learning. Be aware that, in nearly every learning situation, students come with some background knowledge that may be closely or distantly related to the new topic at hand. Perhaps it is something they read or saw on the television, in a movie, or on the Internet. What they know may not be accurate, but you can be sure it will have some degree of influence over the new learning. This influence of past learning on new learning is referred to as *transfer*. As Figure 3.1 illustrates, old learning can either help (called positive transfer) or interfere (called negative transfer) with the acquisition of new learning. Transfer is a powerful principle that teachers should never underestimate, especially in this information-dense environment that our students currently experience.

FIGURE 3.1   The diagram illustrates the impact that established old learning (solid arrow) has on fragile new learning (dotted arrow). The old learning can either help or interfere with the acquisition of new learning. This is an example of transfer during learning.

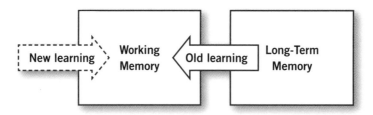

TYPES OF ASSESSMENTS

From the beginning to the end of a unit of instruction, students will progress at different rates, depending on ability, past knowledge about the new learning, interest in the topic, and the methods of instruction. It is important, therefore, to assess what is happening to their brains at various points along the journey to achieving a learning objective. As Figure 3.2 illustrates, teachers should consider a *preassessment* at the beginning of a unit of instruction, formative assessments along the way, and a *summative assessment* at the end. The terms *formative* and *summative* do not describe a *type* of assessment, but *when* that assessment is given and how the results are used. When teachers and students use the assessment results to improve instruction, the assessment is formative. If, on the other hand, the teacher uses the assessment to sum up the learning at the end of a unit of instruction, it is summative. Excellent results on the summative assessment most likely indicate excellent design in the formative assessments. Similarly, poor results on the summative assessment usually reveal poor design in the formative assessments.

Go to www.learningsciences.com/bookresources to download figures and tables.

FIGURE 3.2   The diagram shows the three basic types of assessment: preassessment, formative assessment, and summative assessment.

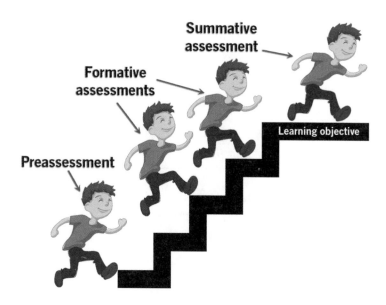

Preassessments

Educators always talk about the lack of time they have in schools to accomplish a burgeoning curriculum. Actually, time is not the enemy; *wasting* time is the enemy. We never can recapture wasted time. This alone is a good reason to determine how much students know about the components of a particular unit of study *before* you teach it. But there certainly are other reasons. Prior learnings are going to have an influence on new learning, so it is important for the teacher to know what those are. The results of the preassessments allow you to:

- decide what instruction should be whole group and what instruction should be with smaller groups

- determine what information students have that could *help* them acquire the new learning (positive transfer)

- discover what students know that could *interfere* with acquiring the new learning (negative transfer)

- identify potential problems that could cause students difficulty with the new learning

- determine how to differentiate instruction

- avoid teaching material the students already know

Go to www.learningsciences.com/bookresources to download figures and tables.

- plan more interesting, engaging, and brain-friendly strategies

- use your planning time more efficiently

- identify students who already have extensive knowledge of the new learning and use them as experts during group instruction

- look for ways to challenge students appropriately

In schools, time is not the enemy;
wasting time is the enemy.

Preassessments can take many forms. Select the form that is appropriate for the age group, the learning objectives, and the time available. They should never be graded. Here are some examples of preassessment formats:

- entrance or exit cards

- individual student interviews

- short written survey about what students already know, understand, and can do

- interest survey

- open-ended questioning in whole-class or group arrangement

- traditional written test

- student demonstration of knowledge and skills

- games based on the topic

- artwork related to topic

- concept maps

- teacher observations

To prepare a preassessment, the teacher should address these questions:

- How long should the preassessment take to complete?

- What should be the form of the preassessment?

- How soon before I start the unit of study should I administer the preassessment?

- Will I have enough time after I see the results of the preassessment to prepare and/or modify my instructional plan?

After answering these questions, the teacher selects the questions for the preassessment. These questions serve to find out any prior knowledge, understandings, and skills the students have that will help them acquire the new learning. If possible, the teacher should consider giving students choices of what method or medium they will use to provide their answers. This approach takes into account the different learning preferences that students have. Not only do students have preferences in how they learn, but they also have preferences in how to show teachers what they have learned. Some may favor taking a paper-and-pencil test or writing an essay; others may prefer drawing a picture or creating lyrics for a song.

Introducing the Preassessment

Preassessments should determine
what is in the students' long-term memories,
not what they find from another source.

The teacher tells the students that they will be starting a unit on [subject]. She says she would like to know what the students already have learned about [subject], either in school or somewhere else. To do that, she will distribute a sheet of paper with some questions that she would like the students to answer. Note: Remember, the teacher is trying to determine what is stored in each student's brain relating to the subject. Asking the students to answer now in class ensures that whatever responses the teacher gets from the students come from their long-term memories because that is the only source of information they have at this moment and in this situation. If the students are allowed to take the preassessment home, they can write down information they find on the Internet (or some other source) rather than what is in their long-term memory. It is also important to tell the students that this activity will *not* be graded. It is intended purely as a guide to help the teacher plan the appropriate instructional strategies for the unit.

Some possible questions on the sheet for different lessons could be:

Example 1. Elementary Fourth-Grade Science Lesson on Rocks and Minerals

- How is a rock different from a plant?

- What are some different ways that rocks are formed?

- Scientists classify rocks into three types. Can you name any of these types?

- What is a mineral, and can you name two examples of minerals?

- Can you think of some objects in the classroom that are made from minerals?

- How do we measure how hard a mineral is?

Example 2. Middle School US History Lesson on the Declaration of Independence

- Why was the Declaration of Independence written?

- Who was the main author of the Declaration of Independence?

- About when was the Declaration of Independence signed?

- In what city was the Declaration of Independence signed?

- Why was signing the Declaration of Independence a courageous act?

Example 3. High School Algebra I Lesson on Introducing Quadratic Equations

- Solve this linear equation: $3x + 4 = 13$

- Factor the following expression: $2x^2 + 3x - 5$

- Did you have any difficulty when answering either of these problems? If so, what was it?

When reviewing the students' responses, the teacher might consider using a simple class matrix that helps identify what individual students know and don't know. Table 3.1 is an example of a partial matrix used with Example 1 on rocks and minerals.

TABLE 3.1 Example of Matrix of Responses to Preassessment

Rocks and Minerals						
Student's Name Grade 4	Rock vs. plant	Ways rocks are formed	3 types of rocks + examples	Mineral + examples	Classroom objects = minerals	Measure hardness of mineral
Bill	Y	Y	N	N	N	N
Kathy	Y	Y	Y	Y	Y	N
Dan	N	N	N	N	N	N
Maria	Y	Y	Y	N	N	N
Christine	Y	N	N	N	N	N
Melinda	Y	N	N	N	N	N
Peter	Y	N	N	N	N	N
Jaime	Y	N	N	N	N	N
Etc.						

Go to www.learningsciences.com/bookresources to download figures and tables.

A review of these partial results indicates that Kathy is already quite knowledgeable about rocks and minerals and that she would be helpful as a peer tutor. Bill and Maria have some basic knowledge of rocks but know little about minerals. Everyone but Dan understands the differences between a rock and a plant, so the teacher does not need to teach this in a whole-class format, thereby saving some time. She can work with Dan individually and ask one of the other students to help, either one-on-one or in a small group.

Table 3.2 shows a similar partial matrix used with Example 3 on introducing quadratic equations.

TABLE 3.2   Example of Matrix of Responses to Preassessment

Introducing Quadratic Equations			
Student's Name Grade 9	Solved equation correctly	Factored expression correctly	Problems + types
Bob	Y	Y	None
Jasmine	Y	N	Forgot how to factor
Drew	N	N	Never had this before
Melissa	Y	Y	None
Wanda	Y	Y	None
Christie	N	N	Had this but can't remember how to do it
Ahmed	Y	Y	None
Joshua	Y	Y	None
Etc.			

Assuming these results are representative of the entire class, how should the teacher plan the first lesson on introducing quadratic equations? Should the teacher review, in a whole-class format, solving linear equations and factoring? Should the teacher set up cooperative learning groups? How should the teacher differentiate instruction, if that would be more efficient? Are there any other ideas for the teacher to consider? Preassessments can provide a wealth of information *before* the teacher plunges into a new topic. They give the teacher the opportunity to plan more interesting challenges for those students who already know some of the material, allowing the teacher more time to spend with those who need to learn the new information and skills.

Go to www.learningsciences.com/bookresources to download figures and tables.

Preassessment of Mindsets

We discussed in Chapter 2 the significant influence that fixed and growth mindsets have over the teaching and learning process. Teachers may get valuable insights into their students by assessing their mindsets early in the school year to determine which have fixed and which have growth mindsets. Some students, of course, will be borderline. The assessment can be a short list of questions with responses on a Likert-type scale, such as "Strongly Agree (SA)," "Agree (A)," "Disagree (D)," and "Strongly Disagree (SD)." Adjust the wording of the questions to be age-appropriate, and include the following ideas:

- You can't really do much to change your intelligence.
  (SA and A = fixed; D and SD = growth)

- Learning new things does not mean you can change your intelligence.
  (SA and A = fixed; D and SD = growth)

- You can change your basic intelligence.
  (SA and A = growth; D and SD = fixed)

- You can't do much to change your amount of talent.
  (SA and A = fixed; D and SD = growth)

- You can change your level of talent.
  (SA and A = growth; D and SD = fixed)

- You like challenges when learning, and you keep trying until you solve a problem.
  (SA and A = growth; D and SD = fixed)

- You recognize that there are some learning challenges you cannot overcome.
  (SA and A = fixed; D and SD = growth)

Be sure to remind students that there are no right or wrong answers. This assessment is appropriate for students ten years of age and older.

Formative Assessments

As the students progress through a unit of study, their cerebral networks are continually reforming as they consolidate new information and skills for future use. This is a complex and not necessarily exact process. The brain can encode misinformation, misunderstandings, and incorrectly learned skills into long-term memory almost as easily as accurate information, understandings, and correctly learned skills. Therefore, teachers should make assessments along the journey through the unit of study to determine whether the students' brains are encoding correct knowledge and not encoding incorrect knowledge, understandings, and skills (see Figure 3.2). Such assessments are called *formative assessments.*

Some researchers refer to formative assessments as assessments *for* learning and assessments *as* learning (Earl, 2013). Assessment *for* learning is teacher centered. It refers to teachers using their personal knowledge of their students' interests and needs to determine the next steps in instruction. Assessment *as* learning is student centered. It refers to students using the results of formative assessments to see their progress toward mastering the learning objectives. These results also help students monitor their own learning and make the modifications needed to successfully improve their performance and reach their learning goals. Formative assessments are like a pilot's course corrections on a long trip. Changes in weather, wind speed, and direction can significantly alter the flight's intended path unless the pilot monitors the instrument panel and makes the necessary adjustments to stay on course. The results of formative assessments allow the students and teacher to determine what corrections they need to make, if any, to stay on course toward achieving the desired learning outcomes.

Types of Brain-Friendly Formative Assessments

Many of the examples of preassessment strategies mentioned earlier in this chapter are also appropriate as effective uses of formative assessments. These ongoing strategies should be keyed to clear learning objectives and the sequence of steps (understandings) that the learner will need to acquire to move along the learning continuum and achieve the final unit objectives (see Figure 3.2). Effective formative assessments give the teacher the information needed to make adjustments in the teaching-learning process while it is still taking place. Here are a few examples of brain-friendly formative assessment strategies:

- **Exit cards.** At the end of the lesson, the teacher poses one or more questions directly related to the day's topic to assess each student's understanding of the lesson's key concepts. This is an effective strategy because, as we discussed in Chapter 2, the student's brain needs to determine sense and meaning to increase the likelihood that the new learning will be stored. Exit cards provide the learner's brain another opportunity to review what it learned, relate the new learning to what the student already knows, and increase the chances that sense and meaning will emerge. This activity reinforces a principle of learning called *closure.* It should take no more than five minutes to complete, and the teacher collects the cards as the students leave the classroom.

- **3-2-1 exit cards.** With this exit card, the students respond to a series of questions requiring three answers, then two, and finally one. For example, a common 3-2-1 exit card asks the students to complete the following: "Three things I learned today are . . . ; two things I found interesting are . . . ; and one question I still have is . . ." Modify the card to suit the learning situation. It possesses all the brain-friendly qualities as the exit card described above.

- **Admit cards.** These are exactly like exit cards but are distributed prior to the beginning of the lesson. The teacher asks the students to reflect on what they learned during the previous day's lesson and/or from their homework. Here again, we have a brain-friendly device because it asks students to recall what they learned from the previous day. Their responses will tell the teacher what learnings the brain likely encoded into long-term memory—important information, indeed. This activity should also take just a few minutes to complete. However, the teacher may want to review a sampling of responses in class to determine what the students remember from the previous day's lesson and/or homework.

- **Graphic organizers.** These are visual models that students complete on a topic or question the teacher poses. Examples are Venn diagrams, What I know (K)–What I want to know (W)–What I learned (L) (known as KWL) charts, mind maps, brainstorming webs, chain-of-events charts, and show-my-thinking charts (see Figure 3.3). All of these organizers are brain friendly because they incorporate the power of visual learning. They help students visualize relationships between and among content and concepts, and reflect on their own understandings and thinking processes.

FIGURE 3.3 Examples of graphic organizers, from left to right: Venn diagram, KWL chart, mind map, and chain-of-events chart.

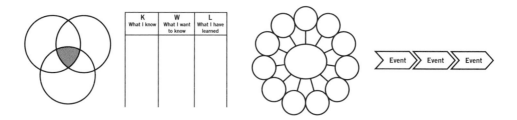

Go to www.learningsciences.com/bookresources to download figures and tables.

- **Learning logs.** Another variation of closure, these logs are journals that students keep to reflect on what they are learning—to look for sense and meaning. The students record the process they use to learn something new and write down any questions they need to have clarified. This allows students to make connections to what they have learned, set learning goals, and reflect on their progress. As a bonus, the very act of writing about their thinking helps students become deeper thinkers as well as better writers. Logs are also an effective way to differentiate instruction. Teachers can monitor each student's progress toward mastery of the learning goals, deliver specific feedback on what the student is doing well and what needs to be improved, and make adjustments accordingly.

- **Questioning.** Brain-friendly questioning goes beyond remembering and understanding—the bottom two levels of Bloom's taxonomy. Recall and mere explanation do not challenge the brain and are usually a reflexive action with little or no reflective thought. Asking questions that require higher-order thinking, however, activates more regions of the learner's brain, calling on neural networks to interact and supply the required answer or alternative solutions (see Figure 3.4). This cerebral challenge motivates and engages the students as they try to piece together the appropriate response (Jauk, Benedek & Neubauer, 2012).

FIGURE 3.4   Note the difference in activity between the brain scan on the left, which is simply recalling information, and the activity in the scan on the right, which is responding to a complex question.

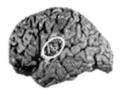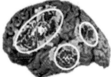

Questioning for deeper thinking. The ultimate goal of education is to produce educated citizens—namely, those who can reflect, reason, and make sensible decisions. Teachers, then, should help students enhance their thinking skills and go beyond mere rote knowledge and basic understandings. The rapidly changing technological world demands certain skills for students to be successful. The knowledge explosion, for example, will require students to judge the validity and credibility of sources and identify the motivation and assumptions behind different points of view.

Questioning provides teachers with a powerful method for assessing the level of students' thinking and to make the necessary instructional adjustments to foster higher-order thinking. Higher-order thinking is generally considered to consist of three components: transfer, critical thinking, and problem solving (Brookhart, 2010). There is substantial

Go to www.learningsciences.com/bookresources to download figures and tables.

research evidence to show a positive relationship between higher-order thinking and student achievement. A meta-analysis of twenty-nine individual studies (twenty from secondary schools) found that students who were regularly involved in higher-order thinking activities showed improved classroom engagement and greater academic achievement (Higgins, Hall, Baumfield & Moseley, 2005). The average effect size of 0.62 on cognitive outcomes for the twenty-nine studies was impressive for an educational intervention. Improvements were also found in students' attitudes and motivation for learning. This should be no surprise. The human brain is designed for problem solving. Asking it to do menial cognitive chores is hardly stimulating. However, give the brain a meaningful and developmentally appropriate challenge, and it is ready for action!

Teachers inadvertently increase the difficulty of a question when they intended to increase its complexity.

Cognitive models. We discussed in Chapter 2 the basics of Bloom's revised taxonomy of the cognitive domain. It has certainly stood the test of time, and it has been a valuable model for designing questions that raise the level of student thinking. Teachers who are comfortable with this model should remember that the levels describe the *complexity* of human thought; that is, the type of thinking at a particular level is more complex than the one below it and less complex than the one above it. However, when creating questions, teachers sometimes inadvertently increase the *difficulty* of a question when they intended to increase the question's complexity. Difficulty describes the amount of neural effort required to answer a question *within* a level of the taxonomy.

A few examples may help clarify the difference between difficulty and complexity. Take this statement: "Name the eight planets in our solar system." This query is at Bloom's Remember level, requiring simple rote recall. What about this one? "Name the eight planets in our solar system in order from the sun." Is this statement more difficult or more complex than the first? The answer: more difficult. It still involves rote recall, but the student needs to exert more neural effort to sort the planets mentally in the correct sequence from the sun. The student's brain is working harder but not practicing any higher level of thinking.

Now consider this question: "After so many years, was it really necessary to downgrade Pluto from a major planet to a dwarf planet? Why or why not?" Answering this more challenging question requires understanding the similarities and differences between a major planet and a dwarf planet (Analysis level), and then rendering an opinion as to whether the downgrading was necessary and why (Evaluation level). Obviously, this question is more complex than just naming the planets. When teachers take care to design questions that are more complex, they have a brain-friendly formative assessment that is more interesting and engaging as well as excellent practice in higher-order thinking.

Some teachers find Bloom's six levels of thinking complexity too cumbersome to use for formative assessment, given the other daily demands they face. Norman Webb designed a simpler four-step model of critical thinking, called the depth-of-knowledge (DOK) scale, that combines Bloom's six levels into four (Webb, 1997, 1999). Webb's DOK four levels of thinking are the following: *Recall* draws on basic knowledge and rote learning and is a combination of Bloom's Remember and Understand levels. *Basic application of skill/concept* is the same as Bloom's Application level. *Strategic thinking* requires research and some synthesis and essentially combines Bloom's Analysis and Evaluation levels. Finally, *extended thinking* is comparable to Bloom's Create level because it includes originality, innovation, and unique creation. Some educators may find Webb's four DOK levels easier to distinguish and use than Bloom's six levels when designing brain-friendly formative assessment questions.

A study that involved nearly 650 observations in ninth-grade classrooms showed that student engagement increased significantly as the instructional process proceeded through the DOK's four levels (Paige, Sizemore & Neace, 2013). Not only did the *degree* of engagement increase, but also so did the *number* of students engaged in each class. Higher-order thinking activities rescue students' brains from rote memorization, boredom, and disengagement. Rather, it challenges them to be a partner in their learning by using their creative talents to solve problems and put forth ideas of their own. Now that is really brain friendly!

Wait time. Teachers often do not give students adequate time to recall the information they need to answer a question. This habit is easily understood. Teachers feel the need to cover the extensive curriculum, and several seconds of quiet in the classroom seems either like an eternity or a waste of time. But tiny intervals—often just two to three seconds—are not compatible with how the young human brain solves problems.

When asking more complex questions, such as, "What do you think might have happened if . . . ?" and "Given what we know so far, what are some plausible endings to this story?" be sure to allow sufficient wait time before calling on a student—perhaps fifteen to thirty seconds, or more, depending on the grade level and complexity of the question. Responding to questions that require deep thinking takes a lot more time than rote recall. Just look at Figure 3.4 to see the difference in brain activity between retrieving a recall response and dealing with a higher-order response. The brain must first interpret the question, start a search of appropriate long-term storage sites, sift through the recalled information, decide what components to use to answer the question, and then select the vocabulary to respond. Because all brains are wired differently, some students do this faster than others. Calling on the first responder without allowing adequate wait time immediately stops the thinking process in all the other students' brains. Why bother thinking any further if the teacher has already called on Joe or Sally to provide the answer?

In addition to adequate wait time, teachers should consider not asking for volunteers when asking questions. Students raise their hands to *ask* a question, not to *answer* it. Such an approach is not necessarily popular at first because the smart students want to show off how much they know, and the students who generally do not participate now have to be engaged. Because no student knows who will be called on for an answer, many more brains get involved in seeking an answer. The percentage of cognitively engaged students goes up—definitely a desireable learning situation. But is this brain friendly? Would the stress of not knowing who is going to be chosen to give the answer raise the students' anxiety levels and dump too much of a hormone called *cortisol* into the blood, thereby halting higher cognitive processing? We will discuss more about cortisol a little later in this chapter.

For most students, not knowing who will be called on probably won't hinder congnitive processing, as long as the teacher handles wrong answers in a way that preserves the responding student's dignity. Remember that most students do not intentionally give wrong answers, because they do not want to look foolish or dumb in front of their peers. They believe their answers are correct. An incorrect answer can often provide an excellent opportunity for clarifying or reteaching a concept. The teacher could respond to an incorrect answer by saying, "That response would be correct if I had asked you . . . " or "I can see why you responded that way, so let me take a minute to clarify the concept." When the teacher's instructional and mangement styles foster a positive learning environment in the class-room, then threats are greatly reduced, and so are the students' anxiety levels. The question now is how does the teacher decide whom to call on? There are numerous ways to do this. One of the more recent methods is to use a randomizer utility found in many electronic whiteboards and on the Internet.

One more point about questioning used as brain-friendly formative assessment: When presenting new information, if all the students are answering all the questions correctly, one has to ask whether the questions are too easy and thus not challenging the students' brains, the students know the material already and the questioning may be a waste of time, or both. Questions should be challenging so that students learn from their correct answers or mistakes and from the ensuing discussion. Incorrect answers prompt the responding student to reexamine the question and determine which cerebral process went wrong to supply an incorrect answer. Research studies show that making mistakes and finding out how to correct them often results in greater learning than not making mistakes (e.g., Huelser & Metcalfe, 2012; Kornell, Hays & Bjork, 2009).

- **Self-assessment.** Student self-assessment is a brain-friendly technique because it allows students to think through how they study and to judge how well they are progressing toward achieving the learning objectives. However, to accurately self-assess, students need to have a clear understanding of where they are going (analyze their work), how they are doing (evaluate their work), and what more they need to do to achieve the learning goal (create a study plan). In other words, self-assessment requires complex, higher-order thinking. Now we have another reason to teach students how to think critically and assess their study habits, cognitive growth, and intellectual progress.

  Self-assessment works when teachers instruct and coach the students on how to do it properly. A study of third and fourth graders who used criteria set in a rubric to regularly self-assess their writing improved the quality of their writing, compared to those students who merely looked over their work (Andrade, Du & Wang, 2008).

- **Indicator cards.** The students select one of three cards to show their level of understanding of the current topic or skill. One card indicates a clear understanding; another suggests the student understands the objective but lacks clarity on some points. The third card indicates the student does not understand the topic or skill. The cards are placed face up on the desk, and the teacher walks around the room to assess how many students fall in each group. There are numerous themes that can be used with this device. For example, weather theme cards can be labled "sunny," "cloudy," and "foggy." Traffic light cards could be "Green: Going OK," "Yellow: Slowing down," and "Red: Need to stop." This device is brain friendly because it allows students to frankly reveal their level of understanding without embarassment.

- **Peer assessments.** Peer assessments can encourage students to see the classroom as a learning community where peers help each other—with the teacher's guidance—to be successful in their work. Students accept sincere praise and constructive criticism as a means to learn about their strengths and weaknesses. However, the teacher must plan this format very carefully to avoid some students' personal issues from undermining the integrity of the process. We will discuss the student feedback criteria in detail later in this chapter.

About 30 percent of secondary students
say homework is busywork, and 33 percent
of secondary-level teachers agree.

- **Homework.** In recent years, whether teachers should give homework and, if so, how much has become a hotly debated issue. Some countries, such as Denmark and Finland, do not give school homework at all. Proponents say a major advantage of homework is that it extends learning time beyond the school day. Homework is brain friendly when it allows students to practice new learnings that they have not yet mastered to an acceptable level. They should not be solving twenty mathematics problems that are essentially reinforcing the same skill, when five will do. Rather, they should be applying that skill to solve new and more challenging problems. Critics ask whether this is really happening in schools. A MetLife survey found that 30 percent of secondary students said their homework was busywork and not related to what they are learning in school (Amos, 2008). Surprisingly, 33 percent of secondary-level teachers said their homework was mainly busywork assignments. Busywork does not encourage students to perform at their best, assuming they complete the assignment at all. A teacher's chances of getting an accurate picture of what the students have learned based on busywork assignments are poor, indeed.

Because homework is formative assessment, teachers should not grade it. However, the results provide the teacher with valuable information about the level of understanding each student has of the topic's main concepts and skills, and not as a final assessment of what the student learned. If the student did not correctly do the homework assignment, teacher feedback should ensure the student that it is okay to fail, because we learn from our errors. This approach reminds the student that mistakes and persistence are part of the learning process. When treated this way, homework becomes a nonthreatening source of communication between student and teacher about the learning objectives.

The homework assignment should be no shorter or longer than necessary to reveal the student's levels of understandings and skills. Because practice makes permanent, teachers should carefully design homework to reinforce understandings and skills learned in class that day, linking the new learnings to past learnings, thereby enhancing existing cognitive networks and increasing retention.

Research on homework. The case for the value of well-planned homework assignments is compelling. However, in recent years, researchers, parents, and students have begun to question whether the quality and quantity of homework that teachers now assign are of true value. When the student's brain perceives homework as a chore, or worse, a punishment, then practically all its value as a learning tool is lost. Poorly structured homework, in other words, could have a negative effect on student achievement. Fortunately, the research evidence supporting the value of carefully designed homework is substantial. One meta-analysis that synthesized the results of other studies over more than a fifteen-year period found that students in classes with appropriate and targeted homework scored 23 percentile points higher on tests of the knowledge presented in that class than students in classes where homework was not assigned (Cooper, Robinson & Patall, 2006).

Because of continual brain growth and development, homework as formative assessment should have different purposes at various grade levels (Cooper, 2007). In the primary grades, teachers should design homework to encourage positive attitudes and habits, permit appropriate parental involvement, and reinforce those skills the student learned in class that day. In the intermediate grades, teachers should assign homework mainly to improve student progress in achieving the learning objective. In middle school and beyond, teachers should design homework to improve the student's depth of understanding and ability to apply the learning to new situations. Improving grades and standardized test scores is a secondary purpose.

Another area of concern is the amount of time students spend on homework each night. The National Center for Education Statistics' latest survey from 2007 found that the average high school student spends nearly seven hours per week on homework. That number is likely higher now, given the increasing pressures of standardized and high-stakes testing in recent years as well as the proliferation of advanced courses. However, the positive effects of homework on achievement are dependent on how much of the homework the student *completes*, rather than the amount of time *spent* on it. Thus, homework that students find as purposeful, meaningful, engaging, and to the point is more likely to be completed and more likely to have the desired positive effect on student progress, while serving as a constructive tool for formative assessment.

Studies show no clear relationship between parental involvement in homework and improved student achievement.

Parental involvement. Despite common beliefs, research studies show no clear relationship between parental involvement in homework and improved student achievement (Robinson & Harris, 2014). In fact, some studies found that regular parental help with homework resulted in lower achievement, regardless of the students' ethnicity. Surprisingly, researchers found this adverse effect even among students who were performing well in school. Possible explanations include parents reporting that they feel unprepared to help and seldom asking their child whether what they are doing is really helping. Further, parental involvement may cause stress in the child, thereby nullifying any positive effects. Teachers can help parents be positive influences on homework if they receive clear guidelines regarding their role, recognize that they are not content experts, and ask questions of their child to clarify what knowledge, understandings, and skills they have learned.

Grading Formative Assessments

Teachers should rarely, if ever, grade formative assessments. If fact, one can argue that teachers should not grade students while they are learning something new. Once students get grades, they stop learning because they shift their focus to the consequences of the grade. Consequently, the brain's emotional system takes over, and cognitive processing of the new learning wanes, reducing the probability that the student will retain any of the learning. This seems to indicate that schools should give grades as seldom as possible.

If formative assessments are graded, students see them
not as helping activities but as punitive activities.

We use formative assessments to assist students in reviewing their performance and making any necessary adjustments to their learning. If these assessments are graded, then students are likely to see them less as helping undertakings but as potentially punitive activities. Their focus shifts to getting a good grade rather than reflecting on how they learn, on how they study, and on their progress. Some may see graded assessments as "gotcha" exercises and feel threatened, thereby diminishing the positive classroom climate needed to build trust among all the students and the teacher.

Furthermore, graded assessments are seldom brain friendly. It all has to do with cortisol, that hormone we mentioned earlier in this chapter. Numerous research studies have shown that elevated levels of cortisol have a negative impact on various bodily functions, including learning, cognitive function, and memory (e.g., Gärtner, Rohde-Liebenau, Grimm & Bajbouj, 2014; Het, Ramlow & Wolf, 2005). Cortisol is always present in the bloodstream in small amounts. It helps the brain stay alert and attend to tasks. Moderate, conventional stress is normal and can actually help students do well on some assessments. But when students know they are taking an important assessment that will be graded, their anxiety increases. This causes the adrenal glands to pump larger amounts of cortisol in the blood to warn the brain that the body is stressed and to prepare for a response that could ease or eliminate the stress. Then working memory, which should be focusing on responding to the assessment, now shifts to calling on long-term memory to help other regions of the brain deal with the cause of the stress. Frustration sets in as the learner has difficulty recalling assessment-related information from preoccupied long-term memory. Soon the student's brain falls into emotional disarray, worrying about the consequences of not scoring well on the assessment.

Ironically, the brain's ability to remember emotional situations is *enhanced* during the period of elevated cortisol levels (van Ast et al., 2013). Consequently, while the cortisol is *impeding* both the memory recall and the cognitive processing needed to complete the assessment, it is also remembering what a stressful situation this is. So the next time the student is faced with an assessment, the emotional memory of the earlier stressful experience (and perhaps a poor grade, as well) is recalled and—what happens?—cortisol levels start to increase again. We often refer to this unpleasant cycle as test anxiety (see Figure 3.5).

FIGURE 3.5   The diagram illustrates the test anxiety cycle. The frustration and emotions of not doing well on one assessment are recalled and induce stress when taking subsequent assessments.

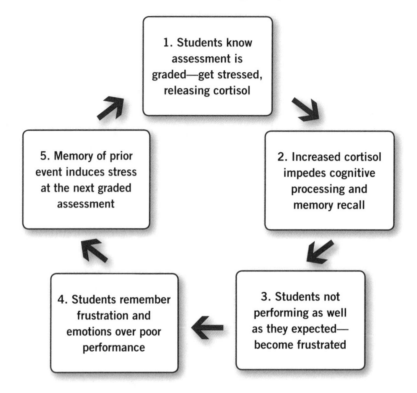

Example of Using Nongraded Assessments: A seventh-grade social studies teacher has discovered the value of formative assessment, especially when used regularly and without the threat of grading. He also knows the value of telling students how their brain learns and remembers and how that information can help them monitor their study habits. Several times a week, he begins a class by distributing a sheet of paper to each student and saying, "Take a few moments to see how well your memory systems are working. Write down three important points you learned and remember from yesterday's lesson on [subject]. Once again, we are just looking at what your brain stored last night." Note that he is asking for this information the next day. He wants to determine what each student retained in long-term memory, and he knows that the transfer of information from working to long-term memory occurs during sleep, as shown in Figure 2.5, which he has shared with the class.

Go to www.learningsciences.com/bookresources to download figures and tables.

After a few minutes, he puts the students in cooperative learning groups of three and gives them this task: "Share your responses and talk about how you remembered each of the learnings you wrote down. Then briefly discuss and write down any memory and study strategies you used that worked for you. You have five minutes, and then I will ask one student to report from each group to the whole class." In this example, the teacher is using the formative assessment as a learning tool, not only to review important content from the curriculum but also to have students reflect on their study methods. He also reminds them to get enough sleep.

Teachers should also not grade homework. Doing so largely defeats its purpose. It becomes a measure of compliance rather than learning. For instance, grading homework rewards those students who learned the information or skill the first time and penalizes those students who are still in the process of learning and who are taking risks. Homework is practice, and students should not have their grade affected because they made errors when practicing new learning. They should get the message that not understanding the learning at the beginning is acceptable. In fact, we often learn from our mistakes, and teachers are here to help students recognize and correct their errors so they can achieve success. Some students may not be able to do their homework because of conditions at home. Domestic violence, alcohol or drugs in the home, a bullying sibling, low self-esteem, and a lack of resources make for a stressful home environment. In any of these circumstances, the student's high stress level is not conducive to focusing on homework assignments.

The Power of Formative Assessments

Decades of studies on the teaching-learning process have revealed a myriad of attributes that influence student achievement. More than 150 influences are in the research literature. Many of them show that the influence under study had positive effects on student achievement. But how powerful were the influences? John Hattie of the University of Melbourne in Australia has synthesized more than nine hundred meta-analyses of fifty thousand research studies involving more than two million students to determine how much impact certain influences have on student achievement (Hattie, 2012). Recall from Chapter 1 that researchers measure the effect of an intervention or influence on a scale called "effect size." It is a useful scale for comparing results on different measures, such as teacher-made tests, student work, and standardized tests.

Hattie's massive study showed that the bar for deciding an influence's effectiveness was set too low if the effect size was set at zero. It turns out that the average effect size for the fifty thousand studies is 0.40 for the influences examined in these studies. (It should be noted that influences not included in Hattie's meta-analysis, and used with different student groups, can have a lower effect size, such as 0.25, to be judged significant.) Influences with effect sizes below 0.40 were below average, and those influences with effect sizes greater than 0.40 were above average. Hattie realized that an attribute needed an effect size of at least 0.40 to be considered effective in improving student achievement. Figure 3.6 shows the effect sizes of a few of the 150 influences included in the synthesis.

> *Feedback is not just from teachers*
> *to students about their learning, but also*
> *from students to teachers about their teaching.*

Note the powerful effect sizes of formative assessments (0.90), teacher clarity (0.75), feedback (0.75), and quality of teaching (0.48). Feedback, by the way, is not just about teachers informing students about their learning, but also the teachers getting student feedback about their teaching. This information can advise teachers about certain instructional strategies that were not effective, or maybe the pace of instruction was too fast or too slow, cooperative learning groups worked better than whole-class discussions, or the learning objectives each day were not clearly stated. The message here is that student achievement is likely to improve when the teaching is clear and of high quality, student–teacher reciprocal feedback is frequent, and formative assessments provide the information to both students and teachers about progress toward accomplishing the learning objectives.

FIGURE 3.6   The chart shows the effect sizes of some common influences on student achievement. The synthesis of fifty thousand studies revealed that an effect size of at least 0.40 was needed for an influence to be effective. Adaped from Hattie (2012).

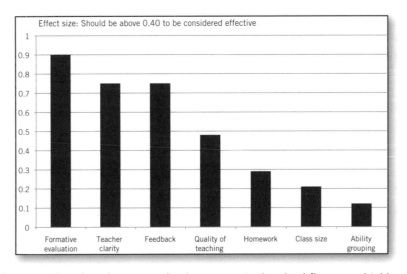

Formative assessment is also cost effective. One extensive study examined the cost effectiveness of twenty-two approaches for boosting student achievement (Yeh, 2011). These approaches included an additional school year, reducing class size, experience level of the teacher, older students tutoring younger students, more rigorous licensing tests for teachers, all-day kindergarten, summer school, and formative assessment. The study found that formative assessment was the most cost-effective approach for improving student achievement when compared to all the other approaches.

Feedback from Teachers

In one study, 70 percent of teachers thought their feedback was clear and useful, but only 45 percent of their students agreed.

As we noted, an effective influence on student achievement that emerged from Hattie's (2012) work is feedback, which, of course, is also a type of formative assessment. But *feedback* is a general term applied to many different activities, some of which are effective and others which are not. One study revealed that teachers focused more on what feedback they gave and not much on what their students understood (Carless, 2006). When asked, about 70 percent of the teachers in this study thought they provided clear feedback that was useful to their students. However, only 45 percent of the students agreed. Other studies reveal that teachers give much of their feedback to the whole class. With this approach, the feedback has little impact because few, if any, students believe that the teacher's comments apply to them. Sometimes the feedback is clear to the teacher but confusing to the students. Even when students do understand the feedback, they often have difficulty relating it to what they are studying (Goldstein, 2006).

As Figure 3.7 illustrates, for feedback to be truly effective for students, studies show it should meet at least these basic criteria (Brookhart, 2012; Chappuis, 2012; Hattie, 2012; Shute, 2008; Tomlinson & Moon, 2013):

- **Emphasize the task, not the learner.** The focus of the feedback should be on the task being assessed, not the learner. This helps the learners realize what they need to do to accomplish the task without worrying about personal shortcomings that could trigger anxiety and unnecessarily raise cortisol levels. For example, "Your conclusions on your lab experiment were well explained, but you should provide more data from the results of your experiment to support them."

FIGURE 3.7   Effective formative assessment should be based on these criteria. Adapted from Brookhart (2012), Chappuis (2012), Hattie (2012), Shute (2008), and Tomlinson & Moon (2013).

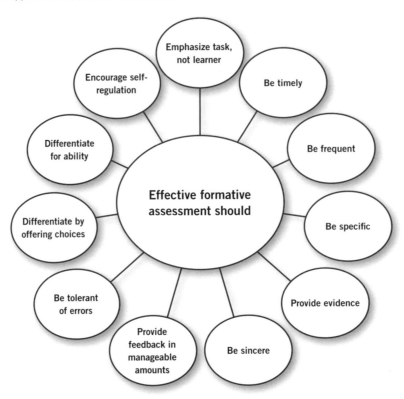

- **Be timely.** Provide feedback as soon as the student completes certain stages in the learning task so that the student can decide what changes to make. For example, "Please make the changes I suggested by tomorrow so we can review them."

- **Be frequent.** Because students should know as soon as possible when they are getting off track in accomplishing the learning task, feedback should be frequent. We do not want students practicing incorrect procedures as they may end up remembering and encoding them into long-term memory. For example, "Check back with me as soon as you make these corrections, and we can see how your work is progressing."

- **Be specific.** Statements like, "Nice job" or "You need to work more on that essay" do not provide the information the student needs to know what made it a nice job or what should be corrected. Why does the student need to work more on the essay? What does the student need to do? Be specific so the student knows which components of the learning task were successful, which need improvement, and how to go about doing it. For example, "Your essay is well written, but you should include more evidence to support your recommendations."

Go to www.learningsciences.com/bookresources to download figures and tables.

- **Provide evidence.** It may be helpful to steer the student to a reliable source of information to support your feedback. Students are often more receptive to objective sources. For example, "I made a list of a few reliable Internet sites you should visit that could provide the information and sources you need to complete your project."

- **Be sincere.** Present the feedback in a way that builds trust between you and the student. It is important for the student to believe that you are sincere, you genuinely care about the student's progress, and you trust the student to use the feedback appropriately. For example, "Decide what sources you think might help you improve your project and show them to me before you begin. We want to be sure you are going in the right direction."

- **Provide feedback in manageable amounts.** Avoid giving too many suggestions at one time as this may result in the brain's working memory experiencing cognitive overload. Consequently, the student may not know where to begin and, therefore, not begin at all. For example, "Here are two suggestions for now. After you work on these, we will look at any other areas you may need to modify to successfully complete the learning objective."

- **Be tolerant of errors.** We often learn a lot from our mistakes. Welcome errors as a normal part of learning new information and skills and as a teaching opportunity, so students feel comfortable telling you of their mistakes. For example, "What you did here is a common error that many students make when first learning this task. Let's see how we can use this as an opportunity for you to overcome that error."

- **Differentiate the feedback by offering choices.** Sometimes it is helpful to provide a student with alternatives for accomplishing a learning goal, and let the student choose from among them. For example, "Here are three ways you could proceed to ensure that your project is successful. Please look them over, pick one, and let me know how you intend to follow through."

- **Differentiate the feedback for ability.** Tailor the feedback to be consistent with the student's achievement level. Immediate, corrective, and scaffolded feedback might work better for low-achieving students. For example, "I just noticed that the steps in your procedure are out of sequence. You want to put your step 3 ahead of step 2. Then we will look at the other sequences." Delayed, expedited, and confirming feedback would be more appropriate for high-achieving students.

"Your work yesterday showed that you really understand and can apply this new procedure correctly. You should move on to the next unit, and let's see what you can do with it."

- **Encourage self-regulation.** Although the feedback makes suggestions to the students, it should not do the students' work for them. Instead, the feedback should give the students the opening they need to make the changes on their own behalf. This encourages students to self-monitor and self-regulate their learning. The idea here is for the students' brains to look for patterns and recognize how to detect and correct errors quickly and accurately. For example, "Review whether your arguments in support of the death penalty are consistent with the evidence you presented. Use that review to determine if there is a more convincing way to present your position."

As Figure 3.6 indicates, feedback meeting these criteria is much more likely to have a positive impact on student achievement than other influences, including grades. Equally important is how the student responds to the feedback. Teachers should present the feedback in such a way that the student responds cognitively ("Oh, I see the error here") rather than emotionally ("You're telling me I made a stupid mistake?"), defensively ("That's what you taught us to do!"), or discouragingly ("I'm just not able to do this"). Remember that negative emotions induce stress, prompting a release of higher levels of cortisol in the bloodstream. As a result, the student's brain will focus on the undesirable consequences of doing something perceived as stupid rather than thinking cognitively about the changes that will correct the error. The memory of this perceived embarrassment is likely to make the student more resistant to teacher feedback in the future.

---

*Timely, specific, and individual feedback
can reduce the students' temptation to cheat.*

Cheating. Another benefit of thoughtful, specific, individual, and regular feedback is that it reduces the need for students to cheat on their assignments. When students do not get the feedback they need to improve their work and do not want a poor test score, they may resort to cheating. When teachers move through the curriculum at breakneck speed, students get the impression that the content is not important. Giving priority to timely feedback that allows students to correct and improve their work lowers the temptation to cheat. By seldom grading formative assessments, and emphasizing that learning is more than test scores and rote recall, teachers help students recognize the value of truly understanding what they are learning.

Feedback from Students

Experienced teachers know that sometimes a student's explanation of a concept can be clearer to other students than the teacher's explanation. The explaining student may choose different vocabulary and sentence structure that resonates with more students and may help clarify misunderstandings. Is the same true for student feedback? Would this be a valuable tool for encouraging student motivation, interest, and self-regulation? And would the peer feedback be accurate? Apparently, not very much so.

One extensive study of in-class observations found that up to 80 percent of verbal feedback in some classes came from peers (Nuthall, 2007). However, much of the feedback was incorrect—well intentioned, but incorrect. Of course, teachers should exercise care when using student feedback as it can have both positive and negative effects. Positive effects include boosting the classroom climate, getting a clearer understanding of the learning objective, and acknowledging success. On the other hand, negative effects include degrading the classroom climate, recognizing misunderstandings of the learning objective, and highlighting poor performance. In a classroom where there are mostly positive relations among students, the students will see carefully designed peer feedback—even constructive criticism—as helpful rather than spiteful (Harelli & Hess, 2008).

Effective peer feedback also has to meet as many of the criteria described previously as possible (see Figure 3.6). Of these criteria, perhaps most important is that the feedback focus on the task and not the learner. This distinction may be difficult for middle and high school students to understand without considerable explanation from the teacher. It is quite easy for preadolescents and adolescents to be critical of their peers' personal traits rather than discuss what went wrong, if anything, with carrying out the learning task. To avoid this unpleasant scenario, the teacher can explicitly inform the students that any feedback should focus on just these three components (Gan, 2011):

- **On the task:** If the student's answer was correct, what did the student do well? If incorrect, where did the student go wrong, and what other information does the student need to meet the learning objective?

- **On the process:** If the student's answer is correct, what successful strategies did the student use? If incorrect, what is wrong, and why? What other questions can the student ask about the learning task, and what is the student's understanding of the concepts as they relate to the learning task?

- **On self-regulation:** How can the student monitor the work done and self-check on progress? How can the student evaluate the information provided and reflect on what was learned?

Throughout this procedure, the teacher will need to be a constant and vigilant coach, providing appropriate prompts to ensure that the feedback stays close to the required criteria and far away from personal issues. With practice and consistent coaching, the students should learn to keep their feedback centered on task, process, and self-regulation. Furthermore, reserve this procedure mainly for learning tasks that are sufficiently complex and challenging, so that there are enough components to provide a significant and meaningful review. Such a review also allows the students to discuss and more deeply understand the concepts and tasks involved in completing the learning objective.

Moreover, by conversing about these components, the students are rehearsing and learning them again. Oral rehearsal is a critical strategy for retention of learning. As students hear themselves talk about their learning, their brains are identifying the neural pathways and networks where this learning will be consolidated and stored, thereby increasing the likelihood that it will be accurately retained and efficiently recalled when needed.

After these steps, peers can enrich the feedback process further by asking some open-ended questions of the student (Gan, 2011). Here are some examples:

- What did you do in order to . . . ?

- How did you justify doing . . . ?

- How can you account for . . . ?

- What does all this information have in common?

- How has this learning experience changed your ideas about . . . ?

- What resources did you find most useful?

These questions will get all students thinking about how to go about solving a problem or accomplishing a learning task and how to develop into autonomous learners. They have the opportunity to reflect on the purposes for learning, examine the steps they took toward achieving the learning goal, monitor their own mistakes, and decide which strategies were effective and which need correction. Effective and brain-friendly formative assessments, including peer feedback, have the potential for helping students acquire the knowledge, understanding, and skills to be autonomous learners while learning the content of their individual subject areas.

In addition, apart from the cognitive growth that peer feedback activities support, there is the influence they have on social and emotional growth. The brains of middle school and high school students are still developing understandings about social skills, relationships, and appropriate interactions with their peers. By learning how to properly use and receive peer feedback, the students are also learning how to phrase constructive criticism so it is helpful and not hurtful, how to acknowledge misunderstandings without reproach, how to be better at recognizing and handling errors, and how to accept responsibility for their errors rather than attributing them to other people or conditions.

Checklist for Designing Formative Assessments

We have discussed a number of components that teachers should consider regarding formative assessments. The checklist in Figure 3.8 may be helpful to teachers as they design and administer these types of assessments. Of course, not all formative assessments need to have every component present. However, teachers should at least consider each component and make a conscious decision whether to include it.

FIGURE 3.8   Checklist to help design formative assessment.

For my formative assessments, have I considered . . .	Check here
Is the assessment brain friendly?	
Will the assessment check for deep understanding of the learning objective?	
Does the assessment encourage higher-order thinking?	
Am I allowing enough time for most students to complete the assessment?	
Have I told the students that I will not grade the assessment?	
Is this assessment appropriate for a peer-assessment activity?	
Is this assessment more appropriate for classwork or a homework assignment?	
Have I decided what type of feedback to use for this assessment?	
Do I have differentiated versions of the assessment?	

Now that we have looked at brain-friendly ways to use preassessments and formative assessments, we turn our attention to devising summative assessments in the next chapter.

CHAPTER 4

Designing and Using
Summative Assessments

Not everything that counts can be counted and
not everything that can be counted counts.

—William Bruce Cameron
author of *Informal Sociology*

SUMMATIVE ASSESSMENTS ARE ASSESSMENTS *OF* LEARNING. THEY TEND TO BE MORE formal than formative assessments, and they reveal and sum up how well the student has achieved the learning objectives at the end of a unit of instruction. I do not equate summative assessments with tests, as we have defined these terms in this book, even though summative assessments are often called "unit tests." Item analysis of summative assessments can provide much valuable information about student achievement as well as how much instruction is aligned with curriculum. Recall that tests are defined as those infrequent, high-stakes events that usually provide little or no specific feedback that is useful for instructional decisions.

No additional formal learning is occurring with summative assessments, except for any assignments and projects that the student may still be completing. Depending on how teachers administer them, summative assessments may not be brain friendly, because students know the assessments count toward their grades and that creates considerable stress. However, students will perceive summative assessments as less stressful if the teacher has effectively used formative assessments along the way and the students have been aware of their progress. This is another reason why formative assessments are so important for maintaining a positive learning environment.

The bad reputation attributed to summative
assessments is not because they are used for grading,
but because of the way the grades have been used.

Unlike preassessments and formative assessments, summative assessments are meant to be graded, and their results used, to determine the student's course grade. Because of this importance, teachers should give different types of summative assessments multiple times during the marking period so that they are not an infrequent event. This gives students who may not have done well on one assessment time to improve their work and do better on the next. Without these opportunities to show improvement, many students reluctantly accept summative assessments as unpleasant and unavoidable. It does not have to be that way. The bad reputation attributed to summative assessments is not because they are used for grading, but because of the way the grades have been used. Instead of viewing the grade as a measure of how much a student has mastered the learning objectives, the grade is too often seen as a symbol indicating that student's placement within a group, be it in a class or school.

TYPES OF SUMMATIVE ASSESSMENTS

Summative assessments may be in the form of chapter or unit tests, projects, midterm and final examinations, and research reports. In general, summative assessments that are written tests fall into two major categories: assessment of rote knowledge (recall) and assessment of higher-level thinking skills. Many summative assessments test rote knowledge for a number of reasons. First, they represent the traditional notion that they measure all the cognitive skills associated with intelligence. Second, their multiple-choice or true/false formats make them easy to grade by hand and even easier by machine. But for years, brain-imaging studies have shown that the brain regions engaged to process rote recall are fewer and different from those involved when processing higher-level mental operations, such as complex problem solving and creativity.

Closed-ended questions that have only one answer evoke rote recall of information, a process referred to as *convergent thinking*. This recall process often occurs with little participation of the brain's prefrontal cortex, where the executive functions are located. After students memorize lists of important dates, names, and definitions, little mental effort is required to retrieve that information from long-term memory just long enough to answer the closed-ended question. The level of complexity is at the lowest levels of Bloom's taxonomy.

The strong emphasis on convergent assessment may be conditioning the young brain for rote recall processing at the expense of higher-level executive functions.

Open-ended questions, on the other hand, have multiple solutions, generating higher-level cerebral processing because no single answer is apparent. This is *divergent thinking*. Look back at Figure 3.4 to see the different brain regions that are most activated during rote recall and during higher-level processing. Notice how much more of the brain is involved when challenged by open-ended questions (Jauk, Benedek & Neubauer, 2012). Many states start high-stakes summative testing as early as the fourth grade. This strong emphasis on convergent assessment, especially with preadolescent students, may be conditioning the brain for rote recall processing at the expense of developing higher-level executive functions (Delis et al., 2007). We recognize that rote skills, such as vocabulary, reading, and mathematics, are important cognitive abilities for achievement in all aspects of life, including academic advancement and career success. But school-age children and adolescents should also receive regular assessments of their higher-level executive functions.

Modern society needs professionals with particular strengths in abstract and creative thinking skills to solve our most serious problems. Yet, the lack of using assessment tools in schools to assess these higher-level cognitive skills hinders our ability to steer the best students into these careers. When we include assessments of higher-level thinking, we can more accurately identify students' weaknesses and strengths in executive functions and guide them into the educational programs and career paths that best fit their needs and abilities. A further argument for assessing higher-level thinking relates to our previous discussion of testing and stress. We explained that in a stressful testing situation, the brain suppresses the recall of information to answer test questions to decide how to respond to the stress. It turns out that this suppression is considerably easier when just a limited region of the brain is involved, as is the case when a student responds to closed-ended questions (see the brain diagram on the left in Figure 3.4). Conversely, the suppression is far more difficult when many brain regions are involved, as is the case when a student responds to an open-ended question (see the brain diagram on the right in Figure 3.4).

Summative assessments containing questions
requiring higher-level thinking and emotional
processing actually reduce the effects of stress.

Now here is a surprise. Brain-imaging studies show that when positive emotions are involved, it becomes even more difficult for the cortisol blockade to take full effect (Buchanan & Tranel, 2008; Sandi & Pinelo-Nava, 2007; Wolf, 2009). This may be because the presence of *endorphins*, brain chemicals associated with positive feelings, further dampens the effects of cortisol. In other words, although summative assessments produce stress, the *effects* of the stress may be reduced when these assessments contain more questions requiring higher-level thinking and emotional processing—thereby involving more brain areas—than questions evoking just rote recall.

Apart from written tests, summative assessments may also be in the form of portfolios, extended projects, and performance tasks related to the learning objectives. Because these are often open-ended assignments, teachers should make clear to their students that the assessment must show what the student knows, understands, and can do regarding the learning objectives and that the student can complete it within the required time. When using standards-based performance assessments, it is better not to assess all the learning objectives at the close of instruction, as this may cause students to see this assessment as a high-stakes activity, leading to anxiety and potentially poorer performance. Summative assessments are less stressful if the teacher seamlessly integrates them throughout the period of instruction.

A common adage in schools is "What gets tested gets taught." If so, then including higher-level summative assessments will encourage teachers to work on higher-level thinking skills with their students. Studies show that teachers can formally introduce these skills as early as third grade (even earlier with some students) and that students who are taught these skills perform better on higher-order thinking assessments than their peers who are not (Bowen, Greene & Kisida, 2014; Burke & Williams, 2008; Glassner & Schwarz, 2007).

Rubrics

Rubrics are a type of spreadsheet that contains a set of standards or expectations that the student can meet at different levels, say from "expert" to "needs work." Research studies indicate that consistent use of rubrics in both formative and summative assessment improves student learning and self-regulation (e.g., Panadero & Jonsson, 2013). The criteria should assess student effort, ability, and performance. Ideally, the teacher and the students should develop the rubric together by writing down exactly what is expected to complete the project. Although the teacher has the last word, allowing students to have input helps them own the learning and the defining criteria. Students receive the completed rubric before they begin a unit or project. When used as summative assessments, rubrics should be task-specific so that the students know what they are expected to know, understand, and be able to do when they complete the learning task. Rubrics are brain friendly because expectations for students are clear at the outset, thereby lowering anxiety.

There are numerous Internet sites with suggestions for rubrics in all subject areas and grade levels. Here are the basic steps for constructing a rubric:

1. Give the rubric a descriptive title.

2. Choose a format. The most common is a matrix, but a list format is also used.

3. Select a rating system. Common ones include four categories: Numbers 1 (lowest) to 4 (highest); Strong, Proficient, Basic, Weak; Advanced, Proficient, Basic, Needs Improvement; Excellent, Nearing Standard, Improving, Emerging; Exemplary, Accomplished, Developing, Beginning.

4. Decide on the criteria or elements the assignment will evaluate for this particular learning unit/project. Examples of criteria include Purpose, Content, Format, Research/References, Organization, Grammar, Spelling, Description, and Conclusion.

5. Decide what attributes will define the various categories of each criterion or element. For instance, what attributes will define "Exemplary" or "Developing" performance in "Organization" or "Accomplished" performance in "Conclusion"?

Figure 4.1 is an example of a summative assessment rubric for a middle school science class designed to assess the quality of a report on a scientific topic. Most rubrics have four to seven categories, but this one has nine. The teacher's assignment of one to four points for the categories is common, as it helps students recognize how the final score was determined. Teachers can also use rubrics as formative assessments, but they should not score them, for reasons we have already discussed.

CHARACTERISTICS OF QUALITY SUMMATIVE ASSESSMENTS

Recognizing that certain summative assessments are districtwide or state standardized tests, it is worth repeating that we are discussing only those assessments that are teacher designed or selected. To be brain friendly and fair, such assessments should possess certain characteristics so both teachers and students feel assured that the assessments are measuring what they intend to measure. Those characteristics include the following (Tomlinson & Moon, 2013; Wormeli, 2011):

- **The assessment should reflect the learning objectives.** The assessment items should align with and measure the knowledge, understanding, and skills that the students were to learn during the course of the unit of study. In short, no surprises!

FIGURE 4.1   This is an example of a rubric that could be used to assess the completion of a report on a topic in science.

RUBRIC FOR REPORT ON SCIENCE TOPIC

Title of Report:_____

Student's Name:_____

	Excellent 4	Accomplished 3	Developing 2	Needs Work 1	Score
Introduction	Gives a precise description of the report	Gives too much information	Gives some information about report	No information about what is in report	
Research	Background research and other interesting facts	Considerable background research	Evidence of some background research	No evidence of background research	
Purpose/ Problem	Addresses an issue directly related to research findings	Addresses an issue somewhat related to the research	Addresses an issue unrelated to the research	Does not address major issue in the topic	
Procedure	Steps are logical and easy to follow	Most steps understandable, but some lack detail	Some steps understandable; most confusing or lack detail	Not sequential, or confusing or missing steps	
Data & Results	Data are accurate and easy to interpret	Data are accurate but illegible characters	Minor inaccuracies in data	Inaccurate or missing data	
Conclusion	Explanation is logical and addresses most or all questions	Explanation is logical and addresses some questions	Explanation not logical but answers few questions	Explanation is not logical and leaves all questions unanswered	
Presentation	Word processed, neatly bound with illustrations	Legible writing, neatly bound with illustrations	Legible writing but print varies, papers stapled together	Illegible writing, loose pages	
Grammar & Spelling	No grammar or spelling errors	One or two grammar and spelling errors	Some grammar and spelling errors	Frequent grammar and spelling errors	
Timeliness	On time	Up to two days late	Up to one week late	More than one week late	
				TOTAL SCORE	

Go to www.learningsciences.com/bookresources to download figures and tables.

- **The assessment items indicate the importance of each learning objective.** The instruction throughout the unit of study should have emphasized and made clear the important concepts and skills for students to master. Therefore, the assessment should include items that focus on this important content and not trivial information.

- **The scope of knowledge required to respond to the assessment should reflect the scope of knowledge needed to master the learning objectives.** For example, if a learning objective is that students should be able to compare and contrast two concepts, but the classroom instruction focused mainly on simply making lists of attributes, then students would likely not score well on an assessment item that requires comparing and contrasting.

- **The cognitive demands of the assessment should match the cognitive demands required to master the learning objectives.** Recall Bloom's taxonomy in Chapter 2. If the learning objective required students to learn some information through rote practice, then the cognitive demands for students to answer assessment terms should be recall (Bloom's Remember level). For this situation, multiple-choice, true/false, or fill-in-the-blank items would be appropriate. If, however, the learning objective called for analysis of several concepts, the assessment items should require analytic thinking, and the formats for recall would not be appropriate. An open-ended essay-type question would be more suitable.

- **Students should not need to have specialized knowledge, understandings, skills, or resources to respond to assessment items beyond that which is available or taught in class.** The student population today is very diverse with regard to culture, language, degree of readiness, and economic status. Teachers should consider these differences when designing summative assessments and avoid any items that call for resources available to some students but not others. They also need to determine whether students who are English language learners have sufficient English proficiency to interpret accurately what the assessment is asking for.

- **Consider retakes.** The one-shot nature of most summative assessments is hardly brain friendly. Numerous factors may interfere with a student's ability to tell a teacher all he really knows when responding to assessment items. Although it cannot be done every time, allowing students to retake standards-based assessments reflects the teacher's understanding that achieving the learning objectives is the most important goal. If it takes more than one try to do so, then that reflects what happens in the real world. In lengthy assessments, the students would need to retake only the portion they did poorly on. Another benefit of retakes is that students are less tempted to cheat if they know they can have another opportunity to take the test.

GRADING AND SUMMATIVE ASSESSMENTS

Summative assessments are meant to be graded, and therein lies the rub. When I was a high school science teacher years ago, my most distressing and least brain-friendly experiences involved deciding what grade to give each of my students on their report cards. Oh sure, I had marks for their tests, experiment reports, projects, homework, and even class participation. It should have been simple, right? Just average the marks (weighted appropriately), and let mathematical calculations do the rest. But I was always uneasy that unknown factors may have affected a student's performance on summative tests. Maybe some students really knew and understood more than their test results reflected. Maybe some students were really good at memorizing the information to answer the questions on my tests while having little true understanding of the science concepts. I did what I thought was best, occasionally nudging a semester grade upward here and there in response to my "gut" feeling about a particular student. Research studies show that I was not alone in this practice (Randall & Engelhard, 2010). But the nagging feeling that this was an imperfect process never left me.

I think it is safe to assume that many teachers have the same disquieting feelings when they calculate grades for their students. I have never met a teacher who said it was an enjoyable event. But do they also recognize the effects that grades have on student learning? Grades, after all, are used as powerful devices that determine a student's rank in class and school, and probably what happens to them after graduation. Longtime education critic Alfie Kohn has argued for abolishing grades entirely in favor of narrative assessments and detailed descriptions of what the student has accomplished while moving through the school's curriculum (Kohn, 2011). He maintains that grades diminish students' interest in learning, encourage them to perform the easiest possible tasks, and reduce the quality of their thinking. He suggests that the more students focus on how *well* they are learning, the less engaged they are with *what* they are learning. The few research studies in this area tend to support his assertions (e.g., Hattie, 2012; Pulfrey, Buch & Butera, 2011).

Regardless of the negative effects that grades have on learning, they are not going away anytime soon. Colleges, universities, and parents want them, although for different reasons. College admissions officers use grades as one of their primary criteria for sorting applicants. Parents want grades because it gives them a sense of how their children are doing, even though many parents do not understand what the grades really mean. So grading will continue as a messy and uncertain process because it is done by humans who have biases and who make errors. The challenge is to determine what practices are likely to minimize the uncertainty and biases that can undermine the reliability and validity of the grades teachers assign to students. Major ones to consider are discussed here and shown in Figure 4.2 (Earl, 2013; Hattie, 2012; O'Connor, 2011).

- **Avoid competition.** If we use grades to compare students to each other, then the students perceive the end goal as competition rather than personal achievement. Teachers can avoid this perception by using grades that reflect students' success in meeting previously identified criteria and that are not norm based, such as grading on a bell-shaped curve. Remember that the normal distribution on a bell-shaped curve is the result of randomly occurring events where there is *no intervention*. In schools, *instruction* is the intervention, and because of instruction, we would expect to get a larger number of students toward the high end of the curve. Otherwise, a normal distribution would indicate that the instruction had no effect. Effective teachers do not sort and rank students but work to ensure that every student achieves the desired learning objectives. Furthermore, studies show that students achieve more when they perceive the learning climate in the classroom as cooperative rather than competitive (Hattie, 2012).

- **Grades reflect clearly defined learning objectives.** Teachers usually make clear to students what they are expected to know, understand, and be able to do as they progress toward the learning goals of the unit under study. They aim their instruction to establish sense and meaning so students see a purpose in what they are learning. Formative assessments along the way help students determine how well they are doing and whether they need to make adjustments in their learning strategies. As teachers also align the summative assessments to the essential learning objectives, students are less anxious and more likely to focus on the primary principles tying the unit concepts together. It is also important for teachers to use similar summative assessments in similar courses to establish the reliability of the assessment—that is, to ensure that it is measuring what it is intended to measure.

- **Assessments are of high quality.** High-quality, brain-friendly assessments are those that measure precisely what they are designed to measure. For example, if the learning goal is rote *recall* of certain information or steps in a procedure, then a multiple-choice or similar type of assessment is appropriate. On the other hand, to determine if students can *apply* what they have learned to new situations, then a performance assessment or product assessment would be better. Using a previously distributed rubric would be a more helpful grading technique because the criteria are well known, and the results make clear to the student which areas were strong and which may need more work.

- **Keep grades to a minimum.** Imagine a middle school music teacher grading a student piano player after every three measures of play, or a high school football coach grading a player after every pass. It may seem silly, but some teachers do tend to grade students *while* they are learning new information and skills. When I questioned a teacher one time about this practice, he argued that "grades encourage students to do their work." However, as we noted earlier, this practice actually has the opposite effect, because it raises students' levels of anxiety and *discourages* learning. Instead, ungraded formative assessments are used along the way to monitor students' progress and give them the opportunity to improve their work in time for the graded summative assessment.

- **Give grades more weight later in school year.** Some students get off to a rocky start when learning a new subject, so their early cumulative grades may be low. Their neural networks are trying to understand the new learning and looking for ways to link it to what the students already know. If a student's work improves later in the school year, that is an indication of sustained effort and growth, and the teacher should acknowledge it by counting the later grades more heavily than the earlier ones. If the final grade is a simple mathematical averaging of all the quarter grades, then it is not an accurate reflection of how much that student's knowledge, understandings, and skills have improved over the course of the school year.

- **Avoid grading items that are not related to accomplishing the learning objectives.** I must admit to being guilty of this practice as a beginning teacher. I thought that if I graded *everything*, then I would have enough data for my end-of-marking-period calculations. Many of my more experienced colleagues supported this practice, remarking, "You can never have too much data when it comes to grades." Consequently, in addition to tests, I graded homework (quality and timeliness), laboratory reports, special reports (content and neatness), degree of class participation, general behavior, and even—[gasp]— attendance. It was only later, after deep soul searching and a better understanding of the variables that affect learning, did I realize how flawed and regrettably unfair this overextended grading process was.

FIGURE 4.2   Brain-friendly grading should have the attributes shown in this diagram.

Surely, some of the items I graded, such as tests and laboratory reports, reflected student learning, but others were uncertain. For instance, we already discussed how the quality of homework can be affected by unfavorable study conditions at home or by input from other sources. Moreover, what about students who seldom turn in homework but score high on the tests? Should their course grades suffer because of the lack of homework? Students can write (or download) information for special reports without much thought processing or true learning taking place. Finally, all sorts of variables can affect class participation, behavior, and attendance that have nothing at all to do with revealing how much the student has achieved. These variables, then, should not be clouding grades. If teachers bundle all of these items into one grade, then that grade is not a true expression of any one of them. Instead, grades should be a clear indication to the student and other stakeholders of what that student knows, understands, and can do because of working toward the learning objectives.

Some researchers in grading believe that summative grades should reflect not only the students' performance or *product*—that is, what the students know, understand, and can do—but also the mental *processes*

Go to www.learningsciences.com/bookresources to download figures and tables.

and work that they are developing, as well as the *progress* the students have made since the previous marking period (e.g., Guskey, 2006). For instance, a student who already has a good background understanding of the new learning may get a higher grade than a student who knows less but studied more and who may end up with a lower grade. Progress measures how much the student learned.

This so-called 3-P approach (for product, process, and progress) to grading, these researchers maintain, is consistent with the growth mindset that teachers should be encouraging in students. The grade would also be more brain friendly because it reports what type of brain growth and development the students experienced while learning the objectives of this unit of study or course. Although report cards will probably continue to use the standard grading system in the near future, teachers could consider using the 3-P format by describing students' growth in these areas in the report card comment section or attaching a written addendum.

- **The grading process should not be a mystery.** Students should know exactly what teachers consider when establishing grades for summative assessments and marking periods. For example: Will rubrics be used? If so, who will create them? Will there be opportunities for performance assessments? Will students be able to retake a test to show they learned the objective? With this transparency, students understand which learning strategies contribute positively to their growth and further academic achievement. Furthermore, grades should not be the only way that teachers communicate with parents about their children's work and progress. Two-way correspondence between teachers and parents ensures that they all are working together to advance the child's learning and achievement.

What Do Grades Accomplish?

There is very little research evidence to show that grades motivate students to perform better.

Grades seldom accomplish what educators and parents really think they accomplish. There is very little research evidence to support the idea that grades motivate good students to perform better. Rather, the evidence suggests that grades often motivate fixed-mindset students to chase after *more* good grades instead of focusing on what they are really learning. Nor do grades motivate struggling learners who may have decided that working harder at learning does not yield results. High-achieving, growth-mindset students learn because they want to and pay little attention to grades or other incentives (Jeong, 2009). Educators and parents should recognize the limited effect that grades have on actually motivating students to learn, and do so by reducing the extended and often erroneous interpretations that all stakeholders read into grades. Grades should represent an accurate assessment of what products, processes, and progress the student made in achieving the learning objects in a particular content area—no more, and no less.

ASSESSING ENGLISH LANGUAGE LEARNERS (ELLS)

Assessing English language learners (ELLs) is a major challenge because poor English language proficiency may prevent them from demonstrating what they really learn. Think for a moment what the brain of each of these students is confronting. First, the brain is working hard to establish another set of neural networks to process the new language. In the early stages of this activity, it has to make connections between the student's native language and newly constructed English language centers to provide basic translations. The older the student, the more labor intensive this whole procedure becomes because the strongly consolidated native- language networks may hinder the development of the weaker and vulnerable English language networks. Language practice, to be sure, helps strengthen these new connections.

Young ELLs are primarily concerned with learning *conversational* English, which, of course, is quite different from the *academic* English used in classroom instruction. Many schools now give ELLs a native language survey and an English language proficiency test as soon as they enroll in the school. These results help establish a baseline to determine the progress the student is making in English language proficiency. Brain-friendly assessment of ELLs occurs when schools assess the student's proficiency in the English language as one grade and arrange a separate grade after assessing the student's academic achievement either in English or the student's native language.

The procedure for identifying, placing, and assessing ELLs is quite complex and outside the scope of this book. Margo Gottlieb (2006), a specialist in this area, suggests that assessment of ELLs is a component of a system wherein classroom and large-scale assessments flow from both language proficiency and academic content standards. These, in turn, influence curriculum and instruction (see Figure 4.3).

FIGURE 4.3   This illustration shows the components of an assessment system for English language learners.

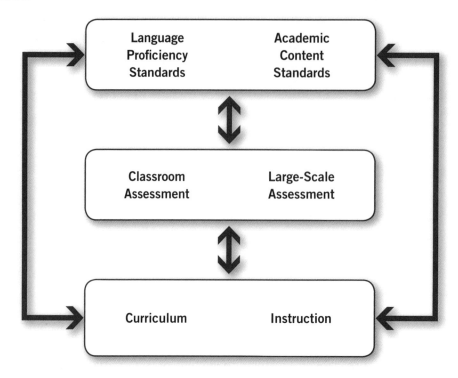

English language learners have many hurdles to confront when entering our schools. In addition to learning both conversational and academic English, they must adjust to socializing with peers from different cultures in a new country. Faced with these challenges, educators need to be sure they design their assessment strategies to show ELLs the road to academic success rather than present obstacles that could lead to failure.

Educators often give little thought to when and where to assess and test students. But there are numerous studies suggesting that the time and place of assessing and testing can affect student performance. We examine these intriguing studies in the next chapter.

CHAPTER 5

Considering When and Where to Assess

Learning and teaching is messy stuff.
It doesn't fit into bubbles. I don't think a simple
pencil-and-paper test is going to capture
what students know and can do.

—Michele Forman
2001 National Teacher of the Year

IN THIS CHAPTER, WE TURN OUR ATTENTION TO WHEN AND WHERE WE SHOULD ASSESS and test students. Although these may seem like minor items, they need to be considered if educators want to do all they can to fairly and accurately allow students to tell and show what they have learned. There is actually a growing body of research suggesting that *when* we assess or test students can make a significant difference in their performance outcomes. Research also indicates that *where* an assessment or test takes place can have an impact on cognitive performance. Let us take a look at both of these situations.

WHEN WE ASSESS CAN MATTER

The question of when to assess actually has two components: When should we assess during the school day? And when should we use high-stakes testing during the school year? Once again, we have to remember that teachers most likely have considerable control over when to do daily assessments, but little control over when the annual high-stakes state and district testing occur. With that in mind, we look first at whether the time of day we assess students can affect their performance.

Assessments during the Day

Research on the *circadian rhythms* and sleep habits of preadolescents and adolescents has focused on their cognitive performance during different times of the day. Studies show that the rhythms and sleep–wake cycle change as children grow into adolescents. These changes are due to natural biological processes and exhibit a preference for adolescents to go to bed later and wake up later in the day than they did as preadolescents (Fischer et al., 2008). If adolescents are awakened before their sleep cycle is complete, they will be sleepy and have difficulty dealing with items requiring their attention as well as substantive cognitive thought. Most of what adolescents do in the morning to prepare for school is routine and therefore automatic, requiring little cognitive processing.

Figure 5.1 shows how the degree of focus for preadolescents and adolescents varies during the course of the day. Note that preadolescents are alert and at a high degree of focus quite soon after they are out of bed. Adolescents, on the other hand, need about an hour more than preadolescents after awakening to attain their full degree of focus. This means that preadolescents would be better prepared for assessments of cognitive performance earlier in the morning rather than later, while adolescents would likely perform better on cognitive assessments later in the morning rather than earlier.

FIGURE 5.1   The graph shows the differences between the degree of focus and time of day for preadolescents (dotted line) and adolescents (solid line). Note that adolescents focus better later in the morning than preadolescents. Focus is an important component of cognitive function and recall during testing. Both age groups experience a drop in their degree of focus past the middle of the day, again later for adolescents.

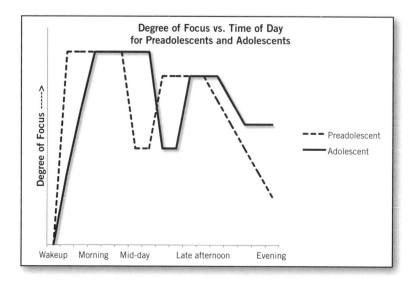

Another part of the chart worth noting is what happens to cognitive focus as the end of the day approaches. For preadolescents, focus tapers off quite rapidly and they become sleepy, but adolescent focus is still significant and sleepiness is delayed (Crowley, Acebo & Carskadon, 2007). This is known as *delayed sleep preference*, and it is not a new discovery. However, it has caught the attention of educators who recognize the irony of school start times. Elementary schools generally start later than high schools. Yet the research we just reviewed tells us that elementary students are ready for cognitive tasks earlier in the morning than high school adolescents. How are circadian rhythms, the sleep–wake cycle, and cognitive performance related? It seems that cognitive performance will vary depending on the duration, difficulty, and type of cognitive processing required for the task (Cote, 2012; Schmidt, Collette, Cajochen & Peigneux, 2007). However, performance on tasks measuring executive functions, such as focus and higher-order cognitive processing (as required for most worthwhile assessments), is optimal for adolescents later in the morning and day (Yoon & Shapiro, 2013).

High school students are getting up earlier than their body clocks expect, thereby accumulating sleep debt.

One major problem stemming from early start times for high schools is that adolescents are getting up earlier than their body clocks expect. As a result, they are accumulating sleep debt during the week (see Figure 5.2). Their sleep–wake clock is biologically set for eight to nine hours of sleep. But if they are getting to sleep around midnight in a school district where the school buses leave at 7:00 a.m., they are getting only about six hours of sleep—a national average according to recent studies (National Sleep Foundation, 2014). The number of sleep hours may be even lower for the growing number of adolescents who use technology under the bedcovers at night—a habit called *vamping*. Early start times for high schools are causing adolescents to attend class, attempt to learn new information and skills, and take assessments long before their cognitive thinking processes are fully operational—not very brain friendly. Likely results are a drop in school attendance, increased tardiness, poorer academic performance than expected, and an increase in behavioral problems (Owens, Belon & Moss, 2010). Many high school teachers have already remarked to me recently about the increasing number of students falling asleep in their first-period classes. This is their bodies' way of trying to complete the preset sleep cycle. One way to avoid these undesirable results is to delay middle and high school start times so that the school schedule is better synchronized with the students' circadian rhythms.

Delaying Middle and High School Start Times

Delaying start times is not easily done, because of the complications that may ensue (Kirby, Maggi & D'Angiulli, 2011). These include rearranging school bus and private transportation schedules; dealing with older siblings taking care of younger family members after school; disrupting schedules for teachers, administrators, and other employees; rescheduling extracurricular and athletic activities; upsetting student participation in nonschool activities; and affecting students' after-school part-time jobs. Despite these difficulties, a number of school districts have dealt with them successfully through collaborative decision making. Studies in districts with delayed start times show some surprising results: adolescents sleep more per night, have better school attendance, and display fewer behavioral problems.

FIGURE 5.2 The graph compares the biologically expected sleep times to the actual average sleep times of adolescents. The difference between these time periods is known as the sleep debt.

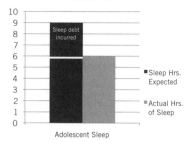

If delaying start times is not practicable, research may offer another option. Recall that the ability of adolescents to focus in the early morning depends on the nature of the cognitive task. Higher-order thinking and processing involving executive functions are not fully functional. However, studies have found that *implicit memory*—memory that does not require conscious attention, such as typing on a keyboard, playing a musical instrument, or riding a bicycle—is not affected by changes in the adolescent's circadian rhythm (Delpouve, Schmitz & Peigneux, 2014; May, Hasher & Foong, 2005). That is why adolescents can go through their get-up and morning routines (implicit memory) with little problem but have difficulty explaining a science experiment (higher-order thinking) shortly thereafter. Altering the middle and high school schedules to place more fluency-based skills, such as orchestra and choral practice and reading, earlier in the day may be more advantageous to students rather than trying to learn new content-heavy and complex material. A major drawback of this arrangement is that it does not address the sleep deprivation problem caused by early start times. Nonetheless, it would diminish somewhat the unfavorable consequences of asking adolescents to attempt demanding cognitive processing when their circadian rhythm and sleep–wake cycle are not synchronized.

Go to www.learningsciences.com/bookresources to download figures and tables.

When to Do Brain-Friendly Assessments during the School Day

Students in the middle and upper elementary grades are ready for significant cognitive processing all during the morning and early afternoon. Assessments done during that time are brain friendly for timing, and the results are likely to reflect academic achievement accurately. To the extent possible, assessments for middle and high school students would be better if given later in the morning. This may not be practicable in first-period classes in schools with early start times. But recognize that the results on assessments may not accurately reflect what these students really know, understand, and can do. Also consider using performance assessments for these early morning classes as they usually require more fluency recall (implicit memory) and less recall of material needing higher-order cognitive processing.

Assessments and High-Stakes Tests during the School Year

We already acknowledged that school districts probably have little or no control over when state education officials decide to administer high-stakes testing. In previous chapters, we discussed that the results of these tests are of little worth for making relevant decisions about the teaching and learning processes. The scores are usually reported in the summer when it is too late to benefit anyone. Nonetheless, educators and parents should at least be aware of the implications that the *timing* of the tests has on student performance.

Most states administer their high-stakes tests between late February and early April. The tests are designed to be summative assessments of how much of the yearlong curriculum the student has achieved. The first problem with this timing is that the tests are given when only 60 to 80 percent of the school year has passed. What if the test contains questions related to curriculum material that was not yet covered in class? State officials have said they account for this when selecting and scoring test items. However, that fact does not alleviate the apprehension students feel when encountering test items about a topic they know nothing about. We already noted in earlier chapters how the supercharged atmosphere in schools leading up to the state tests increases anxiety in principals, teachers, and students, no doubt leading to poorer student performance, thanks again to rising cortisol levels.

A second problem relates to individuals living in the northern and central areas of the country that experience low temperatures and decreased daylight during the winter and early spring months. Between 11 and 25 percent of this population experiences *seasonal affective disorder* (SAD), once known as the winter blues (Byrne & Brainard, 2012) but now recognized as an authentic medical condition (American Psychiatric Association, 2013). Symptoms include decreased energy, irritability, difficulty focusing, and sleeping longer.

SAD is thought to be an evolutionary remnant of ages ago when food was scarce during winter, so a slower metabolism and longer sleep periods meant that less food was needed for energy. Actually, SAD can occur during both the winter and summer seasons, but the former is far more prevalent than the latter. Studies show that individuals with SAD perform worse on tests of visuospatial reasoning, working memory, and verbal memory than those without SAD (Rajajärvi et al., 2010). Students experiencing any or all of these symptoms will not be performing at their best while taking high-stakes tests.

When to Do High-Stakes Brain-Friendly Testing during the School Year

We have already indicated in earlier chapters that high-stakes testing does not improve student learning. There are, of course, other alternatives, such as continual formative assessments, that will yield a more accurate picture of student achievement than one-shot high-stakes testing, and be less stressful on students and teachers. Although educators, parents, and students are voicing plenty of concern about the undesirable consequences of high-stakes testing, the reality is that policy makers remain convinced this is the most effective way to hold schools and educators accountable. Perhaps with continued pressure from educational, parental, and student groups, policy makers will recognize the importance of shifting to a fairer, more research-based, and less stressful system of authentic assessment.

Assessments during the Summer

Ever since the early 1990s, policy makers and school districts have toyed with the idea of extending the school day and school year. Supporters of this notion say it reduces the "learning loss" that occurs over the long summer break, allows deeper coverage of the curriculum, leads to better academic achievement, and makes school buildings more efficient with year-round use. But does it really make a difference? Critics of the idea say that more time does not necessarily translate to improved or increased instruction. Furthermore, student fatigue and boredom can lead to increased absenteeism and dropouts, and there is potential for teacher and administrator burnout. After reviewing the few school districts that have adopted a longer school calendar, researchers seem to agree that extending the school year primarily benefits students most at risk for failure (Patall, Cooper & Allen, 2010). Average achievement scores of all students in these year-round districts went up slightly, but the changes were not significant (e.g., Sims, 2008).

Although extending the school year did not seem to improve student achievement significantly, this does not mean that the summer break has no effect on retention of learning. Hattie's (2012) meta-analysis gives summer vacation an effect size of –0.02, meaning that the long break has a negative effect on student achievement. Nonetheless, in a survey of five hundred teachers by the National Summer Learning Association (NSLA), 66 percent of the teachers said they spent three to four weeks, and 24 percent spent five to six weeks, at the beginning of the school year reteaching course material (NSLA, 2013). Summer learning programs, of course, help reduce this reteaching time.

What Causes Summer Learning Loss?

---

*Many students experience learning loss
in reading and mathematics over the summer,
and these losses add up over time.*

Summer learning loss is the result of the brain competing with itself. *Neuroplasticity* reminds us that the brain constantly restructures itself based on how often we use information and skills, especially in the young brain (Takesian & Hensch, 2013). This restructuring affects our performance so that we get better at the skills and information we practice and worse at what we neglect. If students do not practice over the summer months what they learned during the school year (e.g., reading and mathematics), then the brain thinks it does not need all those neural connections it made for the schoolwork. Consequently, these connections may end up being devoted to activities that the students engage in more often, such a playing sports or video games—a process known as *competitive plasticity*. Many students experience learning loss in reading and mathematics over the summer, and these losses add up over time.

Because the movement to year-round schools is not gaining momentum, and only a small number of students enroll in summer learning programs, assessments over the summer may help reduce learning loss. For example, current-year teachers could coordinate with next-year teachers to design self-assessments that review the material just learned and offer a preview of material to come. They could make the assessments more inviting and brain friendly if they include games and references to Internet sites to get additional information needed to complete the self-assessments.

WHERE WE ASSESS CAN MATTER

It may seem strange to believe that where we assess students might actually make a difference in how well they perform on the assessment or test. But it becomes more believable when we look at research on a not-so-well-known aspect of human memory called *context-dependent memory*. This term refers to improved memory recall of information or an episode when the context where it happened is the same as the context where it was retrieved. Studies have shown that context at the time a memory is encoded can come in various forms, such as external context (the physical or situational surroundings during encoding), internal context (person's state of mind during encoding), and temporal context (the time during encoding). The exact cerebral mechanism by which this occurs is still unclear, but it almost certainly involves the hippocampus—that brain structure responsible for encoding long-term memories—and the prefrontal cortex where higher-order cognitive processing occurs and working memory is located.

Psychological research on external context-dependent memory dates back to the 1930s but did not come to the forefront until the mid-1970s with memory tests of deep-sea divers. Researchers noted that memory recall of deep-sea divers was reduced after they surfaced and got on dry land (Egstrom et al., 1972). They speculated that the change from the underwater environment context to the dry land context inhibited recall, called, appropriately enough, *context-dependent forgetting*. They tested this notion by having divers learn and recall word lists underwater and on land. Results demonstrated that recall of the lists learned underwater was best when in that environment, and similarly for the lists learned on land.

If students are tested in the same environment where they learned the material, they are more likely to have better memory recall than if tested in a different environment.

Other research studies have looked at whether context-dependent memory, similar to that experienced by the deep-sea divers, applied to students learning in a classroom (e.g., Murnane & Phelps, 1994). These studies obtained similar results, but some discovered the good news that context-dependent memory seems to be more powerful with explicit memory (the kind activated during most test taking) than with implicit memory (e.g., Parker, Dagnall & Coyle, 2007). The brain-friendly implication here is that if students are tested in the *same* environment where they originally learned the material, they are more likely to have better memory recall than if tested in a *different* environment.

What if there is no option but to test students in a different environment from where the learning occurred? One major study looked at this scenario. Researchers asked participants to learn a word list in a classroom and note the specific arrangement of desks in rows (Smith, 1984). Later, some of these participants were tested on the word list in a new room (desks arranged in a circle), but were asked to visualize the desk arrangement in the room where they had originally learned the list. They were able to recall as many words as participants who were tested in the original room. Other participants in the new room who did not use this *visualization* technique successfully recalled only about two-thirds of the words recalled by the other groups.

External context-dependent memory is sensitive to other environmental stimuli. Odors can be powerful triggers for memory recall. For example, the odor of a freshly baked apple pie or specific perfume may bring back a flood of memories from long ago. This occurs because smell information, unlike that from all the other senses, travels directly to the brain's limbic (emotional) areas where memories are encoded (Arshamian et al., 2013). Some readers may recall the "peppermint craze" from a few years back. The idea was that if students sucked on peppermint candy while learning something new, the smell of peppermint would make it easier for them to recall that learning later during testing. They would also get an energy boost from the sugar in the candy. Although there is research evidence in external context-dependent memory to support the notion that certain odors assist recall, there is nothing magical about peppermint. It just happens to be a pleasant, strong odor that is easily obtainable from candy or gum. Frankly, a skunk's odor would work just as well—perhaps better because of its shock value—but its availability and acceptability are severely limited.

To sum up, when and where we assess and test students do have an impact on their academic performance. Just taking these conditions into consideration, and fashioning whatever adjustments to place and time of assessments and tests, may make a difference for some students' achievement. And that should be our goal.

In retrospect, we have explored the why, who, what, how, when, and where aspects of assessing student achievement and growth in schools. In the next and final chapter, we summarize some of the major suggestions we have discussed in the book that lead to effective brain-friendly assessments.

CHAPTER 6

Putting It All Together

*The dream begins with a teacher who believes
in you, who tugs and pushes and leads you
to the next plateau, sometimes poking you
with a sharp stick called "truth."*

—Dan Rather
Author, journalist, and former news anchor for the *CBS Evening News*

IF YOU HAVE READ THE PREVIOUS CHAPTERS, THEN YOU KNOW THAT BRAIN-FRIENDLY
assessment is certainly possible—not easy, but possible. It is not easy because
there are federal and state forces ratcheting up educational expectations and
pushing for continued reliance on one-shot, high-stakes testing as the major measure for evaluating the effectiveness of schools and their staffs. Nonetheless, there
is much that teachers can do in their own classrooms to determine student progress toward learning goals with carefully designed, brain-friendly assessments.
We know that assessments can be classified into two major categories: formative
(including preassessments) and summative. Figure 6.1 summarizes the primary
uses of preassessments, formative assessments, and summative assessments. We
also know that formative assessments, including preassessments and performance assessments, and their accompanying feedback have an astonishingly high
impact on student achievement—with impressive effect sizes of 0.90 for formative
assessments and 0.75 for feedback, both well above Hattie's (2012) average effect
size of 0.40. Despite these data, the main focus and funding from policy makers
and state education officials is on summative assessments.

Fortunately, there is some good news: Educators, parents, and students are
beginning to push back, not because they want to abolish testing, but because
they are beginning to realize that the overemphasis on high-stakes testing is
nearly paralyzing schools for months before the test is given. Not much good
comes from that. In the 2014 Phi Delta Kappa/Gallup poll of American attitudes toward public schools, 68 percent of more than one thousand
respondents said that the increase in standardized testing was not helpful to
teachers (Phi Delta Kappa, 2014). Moreover, the practice of using students'

performance results on standardized tests to evaluate teachers was supported by 38 percent but rejected by 61 percent. Although 50 percent of respondents gave their school a grade of A or B, they support neither the prominence of high-stakes testing nor the use of those test scores to evaluate teachers.

FIGURE 6.1   The diagram shows the major uses of the different kinds of assessments that teachers have to determine student progress as well as self-reflect on their teaching effectiveness and the alignment of instruction with curriculum.

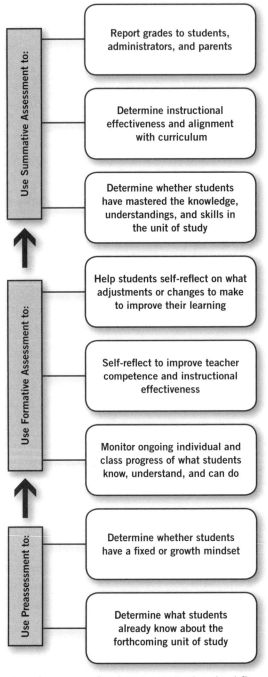

MAKING BRAIN-FRIENDLY ASSESSMENTS POSSIBLE

We need to move from students saying,
"What did you get?" to "What did you learn?"

So where do we go from here? First of all, it is time to recall the nondenominational serenity prayer, setting aside those things we cannot change and working to improve the things we can change. One thing we hope to change is a school culture that focuses too much on grades and not enough on learning. We need to move from students saying, "What did you get?" to "What did you learn?" This occurs when teachers and administrators direct their skills to developing their school as a learning community—a place where *all* participants see themselves as learners.

When it comes to assessing performance, here are some things to consider—gleaned from the previous chapters—that teachers, principals, district leaders, and parents *can* do and perhaps change.

Assessing Students' Performance

We covered this topic in detail in chapters 3 and 4, and here are a few relevant ideas and suggestions worth considering:

- **Commit to authentic formative assessment to improve learning.**
  Schools are first and foremost institutions of teaching and learning. Everything else is secondary. We know enough from educational neuroscience and common sense that students learn best when they are motivated, feel physically and emotionally safe, see relevance in what they are learning, continually know how they are progressing, and are fully engaged in the process. Authentic and brain-friendly formative assessments accomplish all of these things to varying degrees. Of particular importance to the learning process is engagement. Students who are not engaged get discouraged with school. An ongoing Gallup poll of more than six hundred thousand students shows that an average of only 55 percent of students are engaged, highly involved with, and enthusiastic about school (Gallup, 2013). Of the remaining 45 percent, 28 percent are not engaged, and 17 percent are actively disengaged and undermining the instructional process in the classroom (see Figure 6.2). Moreover, the average percent of engaged students steadily declines from fifth to twelfth grade.

FIGURE 6.2   The chart shows the percentage of students who said they were engaged, not engaged, or actively disengaged in school. *Source:* Gallup (2013).

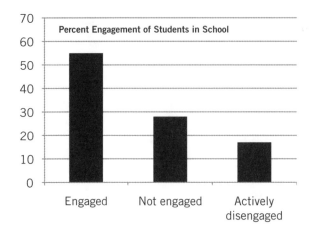

The students in the Gallup survey who strongly agree that their school is "committed to building the strengths of each student," and who have at least one teacher who makes them "feel excited about the future," are thirty times more likely than students who strongly disagree with those statements to be engaged in the classroom—a key predictor of academic success. A 2009 Gallup study of more than seventy-eight thousand students in eighty schools across eight states revealed that a 1-percentage-point increase in a student's score on the engagement index was associated with a 6-point increase in reading achievement and an 8-point increase in mathematics achievement scores (Gallup, 2009). Engagement counts!

Achievement data reveal: Engagement counts!

- **Make learning objectives and criteria for success crystal clear.** Make sure that the learning objectives for the near and long term are stated clearly in age-appropriate language, and clarify any questions students may have about them. Furthermore, students should know what criteria will determine whether they have successfully achieved the objectives. With this brain-friendly approach, formative assessments become practice-for-mastery activities rather than anxiety-producing episodes.

- **Provide continual feedback.** Feedback from formative assessments allows students to mentally process what they need to do to enhance their strengths and correct areas needing improvement. Constructive and specific feedback shows students that the teacher is committed to help all students reach their full potential. It also demonstrates that the teacher *cares* about the students' current and future success, especially when the feedback focuses on a plan for future action rather than a recap of past setbacks. When all teachers in the school adopt this brain-friendly approach, then student engagement is bound to improve, and deeper learning to follow.

- **Remind students that ability is not fixed but incremental.** Teachers should continually be encouraging students to develop or enhance a growth mindset. If students believe they have reached the limit of their abilities when moving toward a learning goal, then motivation diminishes, and they shift their attention to avoiding failure (e.g., Yan, Thai & Bjork, 2014). Here is but another reason for continual, constructive, and specific feedback so students can readily see their progress and growth and recognize that they can achieve learnings they may have thought they never could.

- **Help students become masters of their own learning.** Students are more motivated to learn when they can not only see what progress they are making but also feel they have control over how to improve. When teachers make learning objectives specific and clear, then students start wondering what knowledge and skills they will need to achieve the objectives successfully. To maintain motivation, however, students need to see the learning objectives as attainable and offering some degree of challenge. Some readers may recall Lev Vygotsky's (1978) explanation of this balance as the *zone of proximal development*. To Vygotsky, maximizing learning meant that the task should be a little beyond the learners' reach, posing a reasonable challenge. Teachers, peers, and other knowledgeable people provide the support system to help students bridge the gap between what they already know and can do to what they need to know and do to accomplish the learning task. They monitor the learners' progress along the way through formative assessments, and when the learner has accomplished the first goal, they ratchet up the challenge for the next goal. (Notice that this is precisely how video games entice students to keep playing.) This process is brain friendly because the success of achieving one goal stimulates the reward-processing systems in the emotional area of the brain (Wittman, Bunzeck, Dolan & Düzel, 2007). This stimulation releases *dopamine*, a *neurotransmitter* associated with feelings of pleasure that also increases focus, memory, and motivation (Storm & Tecott, 2005). Gratification from the reward cycle encourages the learner to move on to the next challenge.

When learners know how well they are progressing, motivation stays high.

Teachers should be ready to help students correctly identify the required knowledge and skills and guide them through the learning process. Studies show that when learners know how well they are progressing toward learning goals, whether through teacher feedback or self-assessment, motivation stays high (e.g., Clipa, Ignat & Rusu, 2011; Förster & Souvignier, 2014; Pat-El, Tillema & van Koppen, 2012). We want students to graduate with the skills needed to be self-directed learners—a capital asset for the twenty-first century. And we have not arrived there yet. In the 2014 Phi Delta Kappa / Gallup poll, just 13 percent of respondents agreed that US high school graduates are ready for work, and 31 percent agreed they are ready for college (Phi Delta Kappa, 2014). What are the others ready for?

- **Help students see their peers as resources.** When properly implemented with deliberate instructional support, cooperative learning, peer tutoring, and peer feedback strategies can be powerful forms of brain-friendly formative assessments. Students can often provide insights for one another to recognize errors, add new or corrective information, suggest alternative strategies, and find new ways to solve problems. Hattie's (2012) meta-analysis confirms their significant impact on student achievement: effect size for cooperative learning = 0.59, peer tutoring = 0.55, and peer feedback = 0.53, all significantly above Hattie's overall average of 0.40.

- **Remember students' emotional growth.** Although policy makers are focusing mainly on cognitive student performance, school leaders and teachers should not forget that students grow emotionally as well. In fact, the students' cognitive performance is not likely to be at optimum levels if they do not feel emotionally involved in their learning. Neuroscience makes this sequence clear: emotion drives attention, and attention drives learning. How students *feel* about a topic precedes and determines what they *think* about it. That is why it is important for teachers to set a positive learning environment in the classroom. When students feel positive about their learning environment, powerful chemicals called endorphins are released into the brain. These substances produce a feeling of euphoria and stimulate the brain's frontal lobes, thereby making the learning experience more successful, pleasurable, and likely to be remembered. Students are more willing to engage in a new task and take risks when there is no penalty or embarrassment for making errors (formative assessments are not graded) and where effort, more than ability, is genuinely rewarded.

- **Remember students' social growth.** Schools often do not pay enough attention to the social growth of students. Teachers and administrators can become so focused on covering the curriculum and preparing for high-stakes testing that their students' social development largely goes unnoticed and unattended. Yet the strategies associated with formative assessments offer teachers ample opportunities to assess and encourage their students' social progress. For example, cooperative learning groups, peer tutoring, and peer assessments all involve students interacting with each other. Observing such interactions gives teachers vital clues about individual students' abilities to cooperate rather than compete, listen to others rather than just waiting for their turn to talk, and cope civilly with constructive criticism.

Teachers are brain changers!

As many research studies have shown, teachers are the most important influence over how much and how well students learn. To my mind, the teacher is the only professional whose job is to change the human brain every day. Essentially, teachers are brain changers! If we accept that description, then it makes sense that the more they know about how the brain learns, the more successful they can be at changing it. They need to assess student learning periodically to find out how much change occurred and whether that change is helping students achieve the learning goals.

Conclusion

DESPITE ALL WE SEE AND READ IN THE MEDIA ABOUT THE SORRY STATE OF AMERICA'S schools, it is not all doom and gloom. Granted, US students usually score below those of other countries in international assessments of science, mathematics, and reading. However, it is important to recognize that demography has some effect on the rankings. The United States is dedicated to educating *all* children in a large country with a diverse population and high rate of child poverty. Other countries with high scores tend to have a more homogeneous school population, and they often divert low-performing students to job-training programs rather than academic high schools. Fortunately, most teachers and principals are highly dedicated and hardworking professionals. Apparently, the heavy emphasis in the last decade on high-stakes testing and potentially punitive measures for teacher performance has not done much to improve student achievement. One possible contributing factor may be that pre-service teachers are not getting adequate training in educational measurement theory, assessment types and processes, and modern assessment practices. Several recent studies of teacher education programs show that pre-service teachers typically take a one-semester, three–credit-hour course in educational measurement (e.g., DeLuca & Bellara, 2013; Greenberg & Walsh, 2012). This is hardly enough time for these aspiring teachers to get a deep understanding of the biology of memory and retention or learn how to design and use modern assessment strategies. Furthermore, these courses generally emphasize the importance of data collection (i.e., grades, attendance records, suspension rates, etc.) because of the current era of high accountability. In one major study, of the 179 undergraduate and graduate teacher education programs surveyed, only 21 percent of them adequately covered assessment literacy (Greenberg & Walsh, 2012). Clearly, teacher preparation programs must do a much better job of preparing prospective teachers to understand and use modern, brain-friendly assessments. Otherwise, a whole new crop of beginning teachers will approach assessment as a data-collection process rather than a vital tool for helping students grow and learn in a brain-friendly environment.

There are a few signs that America's schools are slowly getting better. High school graduation rates have climbed in recent years, passing the 80 percent mark in 2013 for the first time in decades. Pushback on overtesting from teachers, administrators, students, and parents is causing federal, state, and local officials to turn to broader measures for assessing school effectiveness. Greater efforts are being made to fairly identify weak and poor teachers and make appropriate decisions on their professional status. Educational neuroscience is telling us more about how the brain learns, so instruction and teacher professional-development programs can be more research based and lead to improvement in student achievement and teacher effectiveness.

No one denies that student progress needs to be regularly assessed. That has been done in many different ways, some more successful than others. The question raised at the opening of this book was whether those assessments could be brain friendly as well as successful. In the preceding pages, I have described ways to make the assessments of students brain friendly by designing and collecting fair and objective evidence of effective performance. By their very nature, schools change slowly, and it takes a great deal of effort to implement changes in how we assess student progress toward instructional goals as well as their emotional and social growth. But change they must to create dynamic systems that can adapt to a rapidly transforming world. As long as the change is continual and improving student achievement, then it is well worth that effort.

Glossary

Automaticity. A condition whereby an individual can perform a skill without conscious deliberation.

Circadian rhythm. The daily pattern of body functions, such as body temperature, breathing, and the sleep–wake cycle.

Classical conditioning. Occurs when a conditioned stimulus prompts an unconditioned response.

Competitive plasticity. The notion that neural circuits are turned over to other uses when an individual does not practice or engage in a specific skill.

Context-dependent memory. The improved memory recall of information or an episode when the context where it happened is the same as the context where it was retrieved.

Convergent thinking. The lower-order thinking process required to recall the answer to a problem that has only one solution.

Declarative memory. Knowledge of facts and events to which we have conscious access.

Delayed sleep preference. A condition caused mainly by a shift in an adolescent's sleep cycle that results in difficulty falling asleep at night and waking up in the morning.

Divergent thinking. The higher-order thinking process required to recall, analyze, and evaluate information to solve problems with multiple solutions.

Dopamine. A neurotransmitter linked to the brain's complex motivation and reward system.

Educational neuroscience. A new field of inquiry that examines how findings from neuroscience can affect the curricular, instructional, and assessment decisions of educational practitioners.

Electroencephalograph (EEG). An instrument that charts fluctuations in the brain's electrical activity via electrodes attached to the scalp.

Endorphins. Opiate-like chemicals in the body that lessen pain and produce pleasant feelings.

Episodic memory. Knowledge of events in our personal history to which we have conscious access.

Effect size (ES). A statistical measure that describes the magnitude of an effect on a group. For studies in education and the social sciences, an effect size of 0.25 or greater is considered significant, although other researchers prefer larger minimum effect sizes.

Formative assessment. An assessment given to students to determine how well they are progressing through a unit of study.

Frontal lobe. The front part of the brain responsible for monitoring higher-order thinking, directing problem solving, and regulating the excesses of the emotional system.

Immediate memory. A temporary memory useful for holding information just a few seconds.

Long-term memory. Areas of the brain where memories are stored up to a lifetime.

Metacognition. A higher-order thinking skill where an individual has conscious control over the cognitive processes while learning.

Mindsets. The beliefs, assumptions, and expectations we possess that direct how we view ourselves and interact with others.

Motivation. The influence of needs and desires on behavior.

Neuron. The basic cell making up the brain and nervous system, consisting of a cell body, a long fiber that transmits impulses, and many shorter fibers that receive them.

Neurotransmitter. One of several dozen chemicals that transmit impulses from one neuron to another.

Nonassociative learning. A learning whereby the individual responds unconsciously to a stimulus.

Nondeclarative memory. Knowledge of cognitive and motor skills to which we have no conscious access, such a riding a bicycle.

Perceptual representation system. A form of nondeclarative memory in which the structure and form of objects and words can be prompted by prior experience.

Neuroplasticity. The ability of the brain to reorganize itself as a result of environmental influences.

Practice. The repetition of a skill to gain speed and accuracy.

Preassessment. An assessment given to students before they begin a unit of study to determine how much they may already know about the forthcoming unit.

Procedural memory. A form of nondeclarative memory that allows the learning of cognitive skills (learning to read) and motor skills (riding a bicycle).

Rubric. A set of standards and expectations, usually in the form of a spreadsheet, used to determine the levels at which a student has acquired learning objectives.

Semantic memory. Knowledge of data and facts that may not be related to any event.

Sensory memory. A temporary memory that monitors the nature and strength of incoming sensory impulses.

Summative assessment. An assessment given to students at the end of a unit of instruction to determine how well they have mastered the learning objectives.

Transfer. The influence that past learning has on new learning, and the degree to which the new learning will be useful to the learner in the future.

Vamping. An activity in which an individual stays up late at night in bed using social media and other technology.

Wait time. The period of teacher silence that follows the posing of a question and before the first student is called upon to respond.

Working memory. A temporary memory where conscious processing of information occurs.

Zone of proximal development. A concept developed by psychologist Lev Vygotsky that describes the range of abilities that a learner can perform with assistance but cannot yet perform independently.

References

ACT. (2012). *The condition of college and career readiness/2012.* Retrieved August 15, 2014 from www.act.org/readiness/2012.

American Psychiatric Association. (2013). *Diagnostic and statistical manual of mental disorders* (5th ed.). Arlington, VA: American Psychiatric Publishing.

Amos, J. (2008). *Do your homework: MetLife survey finds connections between attitude about homework and student achievement, career aspirations.* Washington, DC: Alliance for Excellent Education.

Anderson, L. W. (Ed.), Krathwohl, D. R. (Ed.), Airasian, P. W., Cruikshank, K. A., Mayer, R. E., Pintrich, P. R., Raths, J., & Wittrock, M. C. (2001). *A taxonomy for learning, teaching, and assessing: A revision of Bloom's taxonomy of educational objectives* (Complete edition). New York: Longman.

Andrade, H. L., Du, Y., & Wang, X. (2008). Putting rubrics to the test: The effect of a model, criteria generation, and rubric-referenced self-assessment on elementary students' writing. *Educational Measurement: Issues and Practice, 27*(2), 3–13.

Arshamian, A., Iannilli, E., Gerber, J. C., Willander, J., Persson, J., Seo, H.-S., Hummel, T., & Larsson, M. (2013, January). The functional neuroanatomy of odor evoked autobiographical memories cued by odors and words. *Neuropsychologia, 51*(1), 123–131.

Bloom, B. S. (Ed.), Engelhart, M. D., Furst, E. J., Hill, W. H., & Krathwohl, D. R. (1956). *Taxonomy of educational objectives: The classification of educational goals. Handbook I: Cognitive domain.* New York: David McKay.

Botstein, L. (2014, March 24). The SAT is part hoax, part fraud. *Time, 183*(11), 17.

Bowen, D. H., Greene, J. P., & Kisida, B. (2014, January/February). Learning to think critically: A visual art experiment. *Educational Researcher, 43*(1), 37–44.

Bridgeland, J. M., DiIulio, J. J., & Morison, K. B. (2006). *The silent epidemic: Perspectives of high school dropouts.* Washington, DC: Civic Enterprises.

Brookhart, S. (2010). *How to assess higher-order thinking skills in your classroom.* Alexandria, VA: ASCD.

Brookhart, S. (2012). Preventing feedback fizzle. *Educational Leadership, 70*(1), 25–29.

Buchanan, T. W., & Tranel, D. (2008, February). Stress and emotional memory retrieval: Effects of sex and cortisol response. *Neurobiology of Learning and Memory, 89*, 134–141.

Burke, L. A., & Williams, J. M. (2008, August). Developing young thinkers: An intervention aimed to enhance children's thinking skills. *Thinking Skills and Creativity, 3*, 104–124.

Byrne, B., & Brainard, G. (2012). Seasonal affective disorder. *Therapy in Sleep Medicine*, 695–704.

Carless, D. (2006). Differing perceptions in the feedback process. *Studies in Higher Education, 31*(2), 219–233.

Chappuis, J. (2012). How am I doing? *Educational Leadership, 70*(1), 36–41.

Clipa, O., Ignat, A.-A., & Rusu, P. (2011). Relations of self-assessment accuracy with motivation level and metacognition abilities in pre-service teacher training. *Procedia: Social and Behavioral Sciences, 30*, 883–888.

Cooper, H. (2007). *The battle over homework* (3rd ed.). Thousand Oaks, CA: Corwin.

Cooper, H., Robinson, J. C., & Patall, E. A. (2006). Does homework improve academic achievement? A synthesis of research, 1987–2003. *Review of Educational Research, 76*(1), 1–62.

Cote, K. (2012). Sleep, biological rhythms, and performance. In V. Ramachandran (Ed.), *Encyclopedia of human behavior* (2nd ed., pp. 435–441). Philadelphia: Elsevier.

Cowan, N. (2010). The magical mystery four: How is working memory capacity limited, and why? *Current Directions in Psychological Science, 19,* 51–57.

Crowley, S. J., Acebo, C., & Carskadon, M. A. (2007). Sleep, circadian rhythms, and delayed phase in adolescence. *Sleep Medicine, 8,* 602–612.

Delis, D. C., Lansing, A., Houston, W. S., Wetter, S., Han, S. D., Jacobson, M., Holdnack, J., & Kramer, J. (2007, March). Creativity lost: The importance of testing higher-level executive functions in school-age children and adolescents. *Journal of Psychoeducational Assessment, 25*(1), 29–40.

Delpouve, J., Schmitz, R., & Peigneux, P. (2014, September). Implicit learning is better at subjectively defined non-optimal time of day. *Cortex, 58,* 18–22.

DeLuca, C., & Bellara, A. (2013). The current state of assessment education: Aligning policy, standards, and teacher education curriculum. *Journal of Teacher Education, 64*(4), 356–372.

Diamond, A. (2009, January). All or none hypothesis: A global-default mode that characterizes the brain and mind. *Developmental Psychology, 45,* 130–138.

Dweck, C. S. (2006). *Mindset: The new psychology of success.* New York: Random House.

Earl, L. (2013). *Assessment as learning: Using classroom assessment to maximize student learning* (2nd ed.). Thousand Oaks, CA: Corwin.

Egstrom, G. G., Weltman, G., Baddeley, A. D., Cuccaro, W. J., & Willis, M. A. (1972). *Underwater work performance and work tolerance: Report no. 51.* University of California, Los Angeles: Bio-Technology Laboratory.

Faxon-Mills, S., Hamilton, L. S., Rudnick, M., & Stecher, B. M. (2013). *New assessments, better instruction? Designing assessment systems to promote instructional improvement.* Santa Monica, CA: RAND.

Fischer, F. M., Radosevic-Vidacek, B., Koscec, A., Teixeira, L. R., Moreno, C. R., & Lowden, A. (2008). Internal and external time conflicts in adolescents: Sleep characteristics and interventions. *Mind, Brain, and Education, 2,* 17–23.

Förster, N., & Souvignier, E. (2014, August). Learning progress assessment and goal setting: Effects on reading achievement, reading motivation and reading self-concept. *Learning and Instruction, 32,* 91–100.

Gallup. (2009). *Hope, engagement, and wellbeing as predictors of attendance, credits earned, and GPA in high school freshmen.* Washington, DC: Gallup.

Gallup. (2013). *State of America's schools: The path to winning again in education.* Washington, DC: Gallup.

Gan, J. S. M. (2011). *The effects of prompts and explicit coaching on peer feedback quality.* Unpublished doctoral dissertation, University of Auckland, available online at https://researchspace.auckland.ac.nz/handle/2292/6630.

Gärtner, M., Rohde-Liebenau, L., Grimm, S., & Bajbouj, M. (2014, May). Working memory-related frontal theta activity is decreased under acute stress. *Psychoneuroendocrinology, 43,* 105–113.

Glassner, A., & Schwarz, B. B. (2007, April). What stands and develops between creative and critical thinking? Argumentation? *Thinking Skills and Creativity, 2,* 10–18.

Goldstein, L. (2006). Feedback and revision in second language writing: Contextual, teacher, and student variables. In K. Hyland and F. Hyland (Eds.), *Feedback in second language writing: Contexts and issues* (pp. 185–205). New York: Cambridge University Press.

Gottlieb, M. (2006). *Assessing English language learners.* Thousand Oaks, CA: Corwin.

Greenberg, J., & Walsh, K. (2012, May). *What teacher preparation programs teach about K–12 assessment: A review.* Washington, DC: National Council on Teacher Quality.

Guskey, T. (2006, May). Making high school grades meaningful. *Phi Delta Kappan, 87*(9), 670–675.

Harelli, S., & Hess, U. (2008). When does feedback about success at school hurt? The role of causal attributions. *Social Psychology in Education, 11,* 259–272.

Hattie, J. (2012). *Visible learning for teachers: Maximizing impact on learning.* New York: Routledge.

Het, S., Ramlow, G., & Wolf, O. T. (2005). A meta-analytic review of the effects of acute cortisol administration on human memory. *Psychoneuroendocrinology, 30,* 771–784.

Higgins, S., Hall, E., Baumfield, V., & Moseley, D. (2005). A meta-analysis of the impact of the implementation of thinking skills approaches on pupils. In *Research evidence in education library.* London, England: EPPI-Centre, Social Science Research Unit, Institute of Education, University of London.

Horn, C. V., Zukin, C., Szeltner, M., & Stone, C. (2012). *Left out. Forgotten? Recent high school graduates and the great recession.* Report from the John J. Heldrich Center for Workforce Development. New Brunswick, NJ: Rutgers University.

Huelser, B. J., & Metcalfe, J. (2012). Making related errors facilitates learning, but learners do not know it. *Memory & Cognition, 40*(4), 514–527.

Jacob, B. A. (2002). *Accountability, incentives and behavior: The impact of high-stakes testing in the Chicago Public Schools* (Working Paper 8968). Cambridge, MA: National Bureau of Economic Research.

Jauk, E., Benedek, M., & Neubauer, A. C. (2012, May). Tackling creativity at its roots: Evidence for different patterns of EEG alpha activity related to convergent and divergent modes of task processing. *International Journal of Psychophysiology, 84*(2), 219–225.

Jeong, D. W. (2009, December). Student participation and performance on advanced placement exams: Do state-sponsored incentives make a difference? *Educational Evaluation and Policy Analysis, 31*(4), 346–366.

Johnson, J., Arumi, A. M., & Ott, A. (2006). Is support for standards and testing fading? *Reality Check 2006.* New York: Public Agenda.

Kirby, M., Maggi, S., & D'Angiulli, A. (2011). School start times and the sleep–wake cycle of adolescents: A review and critical evaluation of available evidence. *Educational Researcher, 40*(2), 56–61.

Kohn, A. (2011, November). The case against grades. *Educational Leadership, 69*(3), 28–33.

Kornell, N., Hays, M. J., & Bjork, R. A. (2009). Unsuccessful retrieval attempts enhance subsequent learning. *Journal of Experimental Psychology: Learning, Memory, and Cognition, 35,* 989–998.

Lepi, K. (2013, July 16). *This is why teachers quit.* Available online at http://www.edudemic.com /this-is-why-teachers-quit/.

May, C. P., Hasher, L., & Foong, N. (2005). Implicit memory, age, and time of day: Paradoxical priming effects. *Psychological Sciences, 16,* 96–100.

Mitchell, J. P., Banaji, M. R., & Macrae, C. N. (2005, August). The link between social cognition and self-referential thought in the medial prefrontal cortex. *Journal of Cognitive Neuroscience, 17,* 1306–1315.

Murnane, K., & Phelps, M. P. (1994). When does a different environmental context make a difference in recognition? A global activation model. *Memory & Cognition, 22*(5), 584–590.

National Center for Education Statistics. (2007). *Parent and family involvement in education survey of the national household education surveys program.* Washington, DC: U.S. Department of Education.

National Sleep Foundation. (2014). *2014 sleep in America poll: Sleep in the modern family.* Arlington, VA: Author.

National Summer Learning Association. (2013). *Teachers confirm time wasted due to summer learning loss.* Baltimore, MD: Author.

Nichols, S. L., Glass, G. V., & Berliner, D. C. (2012). High-stakes testing and student achievement: Updated analyses with NAEP data. *Education Policy Analysis Archives, 20*(20). Retrieved July 17, 2014 from http://epaa.asu.edu/ojs/article/view/1048.

Nuthall, G. A. (2007). *The hidden lives of learners.* Wellington: New Zealand Council for Education Research.

O'Connor, K. (2011). *A repair kit for grading: 15 fixes for broken grades* (2nd ed.). Boston: Pearson.

Owens, J. A., Belon, K., & Moss, P. (2010). Impact of delaying school start time on adolescent sleep, mood, and behavior. *Archives of Pediatrics and Adolescent Medicine, 164*, 608–614.

Paige, D. D., Sizemore, J. M., & Neace, W. P. (2013). Working inside the box: Exploring the relationship between student engagement and cognitive rigor. *NASSP Bulletin, 97*(2), 105–123.

Panadero, E., & Jonsson, A. (2013, June). The use of scoring rubrics for formative assessment purposes revisited: A review. *Educational Research Review, 9*, 129–144.

Parker, A., Dagnall, N., & Coyle, A.-M. (2007). Environmental context effects in conceptual explicit and implicit memory. *Memory, 15*(4), 423–434.

Patall, E. A., Cooper, H., & Allen, A. B. (2010, September). Extending the school day or school year: A systematic review of research (1985–2009). *Review of Educational Research, 80*(3), 401–436.

Pat-El, R., Tillema, H., & van Koppen, S. W. M. (2012, August). Effects of formative feedback on intrinsic motivation: Examining ethnic differences. *Learning and Individual Differences, 22*(4), 449–454.

Phi Delta Kappa. (2014). *The 46th annual PDK/Gallup poll of the public's attitudes toward the public schools.* Arlington, VA: Author.

Price, J. M., Colflesh, G. J. H., Cerella, J., & Verhaeghen, P. (2014, May). Making working memory work: The effects of extended practice on focus capacity and the processes of updating, forward access, and random access. *Acta Psychologica, 148*, 19–24.

Pulfrey, C., Buch, D., & Butera, F. (2011). Why grades engender performance-avoidance goals. *Journal of Educational Psychology, 103*(3), 683–700.

Rajajärvi, E., Antila, M., Kieseppä, T., Lönnqvist, J., Tuulio-Henriksson, A., & Partonen, T. (2010, December). The effect of seasons and seasonal variation on neuropsychological test performance in patients with bipolar I disorder and their first-degree relatives. *Journal of Affective Disorders, 127*(1–3), 58–65.

Randall, J., & Engelhard, G. (2010, October). Examining the grading practices of teachers. *Teaching and Teacher Education, 26*(7), 1372–1380.

Robinson, K., & Harris, A. L. (2014). *The broken compass: Parental involvement with children's education.* Cambridge, MA: Harvard University Press.

Rowland, C. (2013, April 8). *Florida law says blind, severely disabled child must be tested.* Orlando, FL: News 13. Retrieved July 17, 2014 from http://mynews13.com/content/news/cfnews13/news/article.html/content/news/articles/bn9/2013/4/7/florida_law_says_bli.html.

Sandi, C., & Pinelo-Nava, M. T. (2007, April). Stress and memory: Behavioral effects and neurobiological mechanisms. *Neural Plasticity, 2007,* 1–20.

Schmidt, C., Collette, F., Cajochen, C., & Peigneux, P. (2007). A time to think: Circadian rhythms in human cognition. *Cognitive Neuropsychology, 24,* 755–789.

Schuette, C. T., Wighting, M. J., Spaulding, L. S., Ponton, M. K., & Betts, A. L. (2010). *Factors that influence teachers' views on standardized tests.* Lynchburg, VA: Liberty University Faculty Publications and Presentations. Paper 157. Retrieved July 20, 2014 from http://digitalcommons.liberty.edu/educ_fac_pubs/157/.

Schwartz, B. L., & Bacon, E. (2008). Metacognitive neuroscience. In J. Dunlosky & R. A. Bjork (Eds.), *Handbook of metamemory and memory* (pp. 355–372). New York: Psychology Press.

Shaw, L. (2013, February 5). Parents joining teachers' test boycott as Garfield High principals give exam. Seattle: *The Seattle Times.* Retrieved July 23, 2014 from http://seattletimes.com/html/localnews/2020294766_garfieldtestxml.html.

Shimamura, A. P. (2008). A neurocognitive approach to metacognitive monitoring and control. In J. Dunlosky & R. A. Bjork (Eds.), *Handbook of metamemory and memory* (pp. 373–390). New York: Psychology Press.

Shute, V. J. (2008). Focus on formative feedback. *Review of Educational Research, 78*(1), 153–189.

Sims, D. P. (2008). Strategic responses to school accountability measures: It's all in the timing. *Economics of Education Review, 27,* 58–68.

Smith, S. M. (1984). Comparison of two techniques for reducing context-dependent forgetting. *Memory and Cognition, 12*(5), 477–482.

Sousa, D. A., & Pilecki, T. (2013). *From STEM to STEAM: Using brain-compatible strategies to integrate the arts.* Thousand Oaks, CA: Corwin.

Sousa, D. A., & Tomlinson, C. A. (2011). *Differentiation and the brain: How neuroscience supports the learner-friendly classroom.* Bloomington, IN: Solution Tree Press.

Sperling, R. A., Howard, B. C., Staley, R., & DuBois, N. (2004). *Educational Research and Evaluation, 10*(2), 117–139.

Statistic Brain. (2014). *High school dropout statistics.* Statistic Brain Institute. Retrieved July 17, 2014 from http://www.statisticbrain.com/high-school-dropout-statistics/.

Storm, E. E., & Tecott, L. H. (2005). Social circuits: Peptidergic regulation of mammalian social behavior. *Neuron, 47,* 483–486.

Strauss, V. (2013a, March 11). Student protests against standardized tests spreading. *The Washington Post.* Retrieved July 23, 2014 from http://www.washingtonpost.com/blogs/answer-sheet/wp/2013/03/11/student-protests-against-standardized-tests-spreading/.

Strauss, V. (2013b, May 2). Hospitalized 4th grader, hooked up to machines, asked to take standardized test. *The Washington Post.* Retrieved July 23, 2014 from http://www.washingtonpost.com/blogs/answer-sheet/wp/2013/05/02/.

Takesian, A. E., & Hensch, T. K. (2013). Balancing plasticity/stability across brain development. *Progress in Brain Research, 207,* 3–34.

Tomlinson, C. A., & Moon, T. R. (2013). *Assessment and student success in a differentiated classroom.* Alexandria, VA: ASCD.

Tompson, T., Benz, J., & Agiesta, J. (2013). *Parents' attitudes on the quality of education in the United States.* Chicago: Center for Public Affairs Research.

Tyrrell, J. (2013, November 16). 248 LI principals join protest against over-testing. *Long Island Newsday.* Retrieved July 20, 2014 from http://www.newsday.com/long-island/suffolk/248-li-principals-join-protest-against-over-testing-1.6449414.

van Ast, V. A., Cornelisse, S., Marin, M.-F., Ackermann, S., Garfinkel, S. N., & Abercrombie, H. C. (2013). Modulatory mechanisms of cortisol effects on emotional learning and memory: Novel perspectives. *Psychoneuroendocrinology, 38*(9), 1874–1882.

Vygotsky, L. S. (1978). *Mind in society: The development of higher psychological processes.* Cambridge, MA: Harvard University Press.

Webb, N. (1997). *Criteria for alignment of expectations and assessments on mathematics and science education* (Research Monograph No. 6). Washington, DC: CCSSO.

Webb, N. (1999). *Alignment of science and mathematics standards and assessments in four states* (Research Monograph No. 18). Washington, DC: CCSSO.

Wittman, B. C., Bunzeck, N., Dolan, R. J., & Düzel, E. (2007). Anticipation of novelty recruits reward system and hippocampus while promoting recollection. *NeuroImage, 38*, 194–202.

Wolf, O. T. (2009, October). Stress and memory in humans: Twelve years of progress? *Brain Research, 1293*, 142–154.

Wormeli, R. (2011, November). Redos and retakes done right. *Educational Leadership, 69*(3), 22–26.

Yan, V. X., Thai, K.-P., & Bjork, R. A. (2014, September). Habits and beliefs that guide self-regulated learning: Do they vary with mindset? *Journal of Applied Research in Memory and Cognition, 3*(3), 140–152.

Yeh, S. S. (2011). *The cost-effectiveness of 22 approaches for raising student achievement.* Charlotte, NC: Information Age.

Yoon, S. Y. R., & Shapiro, C. M. (2013). Chronobiology of sleep: Circadian rhythms, behavior, and performance. *Encyclopedia of Sleep*, 426–434.

Young, A., & Fry, J. D. (2008, May). Metacognitive awareness and academic achievement in college students. *Journal of the Scholarship of Teaching and Learning, 8*(2), 1–10.

Index

Notes

Notes

Notes

Notes

Notes

Notes